P9-AOV-705

The Microeconomics of the Timber Industry

Westview Replica Editions

This book is a Westview Replica Edition. The concept of Replica Editions is a response to the crisis in academic and informational publishing. Library budgets for books have been severely curtailed; economic pressures on the university presses and the few private publishing companies primarily interested in scholarly manuscripts have severely limited the capacity of the industry to properly serve the academic and research communities. Many manuscripts dealing with important subjects, often representing the highest level of scholarship, are today not economically viable publishing projects. Or, if they are accepted for publication, they are often subject to lead times ranging from one to three years. Scholars are understandably frustrated when they realize that their first-class research cannot be published within a reasonable time frame, if at all.

Westview Replica Editions are our practical solution to the problem. The concept is simple. We accept a manuscript in camera-ready form and move it immediately into the production process. The responsibility for textual and copy editing lies with the author or sponsoring organization. If necessary we will advise the author on proper preparation of footnotes and bibliography. We prefer that the manuscript be typed according to our specifications, though it may be acceptable as typed for a dissertation or prepared in some other clearly organized and readable way. The end result is a book produced by lithography and bound in hard covers. Initial edition sizes range from 600 to 800 copies, and a number of recent Replicas are already in second printings. We include among Westview Replica Editions only works of outstanding scholarly quality or of great informational value, and we will continue to exercise our usual editorial standards and quality control.

The Microeconomics of the Timber Industry

David H. Jackson

This book outlines the conditions for a market solution to the historically controversial issue facing timber manufacturers--the specter of market failure in an industry that requires long time spans for production. Linking the specific theory of timber investment with the broad market theory of supply and demand under conditions of perfect competition, the author develops a theory of timber production and supply that considers both short-term market speculation and long-term timber supply, and he defines the marginal conditions requisite to efficient timber production in differing circumstances throughout the life of the stand.

Dr. Jackson then applies his theoretical models to the question of forest taxes. Reporting tax-induced changes in land use, timber supply and price, and the utilization of harvested timber, he describes the competitive advantage of the private firm in an industry composed of a private sector and an untaxed public sector.

David H. Jackson is assistant professor in the school of forestry at the University of Montana and has taught at the University of Alberta.

The Microeconomics of the Timber Industry

David H. Jackson

Westview Press / Boulder, Colorado

A Westview Replica Edition

Published in 1980 in the United States of America by
Westview Press, Inc.
5500 Central Avenue
Boulder, Colorado 80301
Frederick A. Praeger, Publisher

Library of Congress Cataloging in Publication Data
Jackson, David H. 1940-
The microeconomics of the timber industry
(A Westview replica edition)
Bibliography: p.
1. Lumber trade--United States. I. Title.
HD9756.J32 338.4'7674'0973 80-14631
ISBN 0-89158-887-6

Printed and bound in the United States of America

To
Dylan, Alexander, Lucas
and Anna Louise

Contents

Preface

Most of the material in this book was prepared originally as a doctoral thesis of the same title. Shortly after its completion, I was asked to present a paper as a part of a panel at the 1975 Midwestern Forest Economists meeting. The paper which was based on material in this text was entitled "On the Theory of the Timber Firm and the Incidence of Forest Taxes." Just before presentation of my paper, Dick Alston, the panel moderator asked that I "be controversial." In that spirit, I mentioned that the paper which had already been distributed could be retitled; "The Competitive Theory of Timber Production: A Capitalistic Manifesto to Sustained Yield Forestry." Judging by the response, the importance of efficiency is ignored by some leading forest economists.

The model of perfect competition and private market exchange has been extremely useful in economic analysis in terms of describing productive and consumptive efficiency in addition to locating potential sources of market failure. Unlike most forms of commodity production, forests in the United States and indeed in many of the other mixed capitalistic economies of the world are partially owned and managed by agencies of the state. In the U.S., the early impetus for public intervention was a perception or belief that private interests were incapable of maintaining future supplies of timber. In hindsight, fears of private resource exhaustion with respect to timber have not been fulfilled, and in effect, federal timber supplies have been perhaps the least dependable in recent years.

It would be inaccurate to infer that the sole role of the national forests is the production of timber. Since the enactment of the national forests, formal public policy has gradually embraced the objective of managing lands for multiple resource uses. In addition to timber, the forests are directed to produce water, wildlife, grazing and aesthetic (recreation) benefits. Due to a variety of institutional problems, these goods and services may be inefficiently allocated in the marketplace. Whether the national forests are actually solving problems of social inefficiency related to multiple uses or whether current national forest management is itself a form of non-market failure is a point open for conjecture.

In either event, the value of a study of competitive timber production and supply is that it provides a basis for comparison with actual production decisions. Whether timber is the only resource with significant value on a particular piece of land or whether there are potential means of altering decisions in order to increase the calculus of social benefits and costs resulting from other uses, the competitive model of the "exchanged" resource should serve as a familiar point of departure.

The purpose of this book is to provide a better linkage between microeconomic theory and forestry. The intended audience is forest economists, resource economists, graduate students interested in forest management and economics and others interested in an economic framework useful in viewing major public policies. For those who are fairly uninitiated in forestry, there is a more substantive foundation in timber production than the classic textbook case of "when to cut the trees." For those readers who are less familiar with comparative statistics techniques, fairly detailed proofs of the major propositions are developed in the Appendices. Hopefully, this book also presents some basic foundations for further research and analysis.

Acknowledgments

I gratefully acknowledge many contributions in terms of encouragement, assistance and criticism. Kathleen O'Brien Jackson has provided immense help and support. Marion Clawson's encouragement and suggestions led to publication. Jack Weeks made helpful comments as did Gerard Schreuder, Thomas Waggener, Barney Dowdle, James Crutchfield, William Alberts and Joseph O'Leary. My thanks goes to Resources for the Future, Inc. for their financial support in the form of a Dissertation Fellowship Grant. The Department of Forest Science at the University of Alberta and the Montana Forest and Conservation Experiment Station at the University of Montana provided assistance as well. The book would have been impossible without Elaine Gerard's typing and Cindy Romo's graphics talents. However, none of the above are responsible for what I have written.

1 Nature of the Research Problem

THE ECONOMIC QUESTION OF TIMBER SUPPLY

Whether or not timber will be available in the future remains a problem currently investigated in the forestry profession. This question is not new. Pinchot's publication of The Fight for Conservation (Pinchot, 1910), is an early prediction of doom. Pinchot indicated at that time, the United States had already "crossed the verge of a timber famine so severe that its blighting effects will be felt by every household in the land." (Pinchot, p. 15). The basis for Pinchot's prediction was similar to the rationale underlying many contemporary environmental polemics. Using information on the stock of standing timber or timber inventory and the then yearly rate of harvest, he stated that the "...probable duration of our supplies is little more than a single generation." (Pinchot, p. 16). Suffice it to say that almost seventy years later, no timber famine has beset the economy.

More recently, the United States Forest Service has predicted a diminishing availability of timber from the major producing region in the United States by 1980 (the Douglas-fir region). The 1973 report entitled The Outlook for Timber in the United States (U.S. Forest Service, 1973, p. 82), projects a 25 percent reduction in private forest industry roundwood products output from the Pacific Coast region over the period of 1970 to 1980, and a continued further decline to the year 2020. These projections are made on the basis of a constant 1970 level of management activity in the face of projected increases in demand for forest products in the current decade. If these projections are true, their significance is of vast importance. For example, one paper which accepts the projections at face value predicts a 45 percent reduction in forest based employment from 1970 to 2000 in Washington and Oregon alone. Wall (1973, p. 4), calculates a direct employment loss of some 54,000 jobs as a result of the private industry reduction in timber output.

The abiding concern with timber supply both in the historical record and in the context of predictions of future supply warrants an investigation of the economic theory of timber supply. If the above projections are correct, output in the Pacific Northwest will fall in the current decade during a period of projected record demand. Such a result is certainly not in keeping with the general economic theory of of supply.

In addition to the recurring theme of future timber supplies, there is a growing concern with the legal mandates of the National Forest System in terms of their efficiency and social welfare considerations. Waggener (1960), Clawson (1975, 1976), Samuelson (1976), Baden and Stroup (1973, 1975) Walker (1977), and Howe (1979) have expressed a variety of concerns ranging from pure efficiency in timber investment and supply to interresource efficiency in terms of multiple uses. Krutilla and Haigh (1978) recently concluded that the laws governing the national forests do not preclude socially efficient operations.

The abiding interest with timber supply along with the growing awareness of economic aspects of national forest policy suggests a relative weakness with our understanding of forest economics. Currently there is no comprehensive theory of timber production under conditions of perfectly competitive markets. This is not to say that the necessary elements do not exist for the foundations of such a theory. Rather, the elements have not been drawn together in order to examine timber investment, production and supply in a partial equilibrium context. As a result, our basis for responding to major policy issues is wholly inadequate. Furthermore, if multiple uses are in fact forms of externalities in terms of our current institutional environment, a model of efficient timber production should prove to be of enormous assistance in achieving socially efficient allocation of resources.

Demand for any product fluctuates over time. A topic closely related to the future availability of timber is the responsiveness of timber output to short term changes in market conditions. What conditions determine the short term changes in output in the light of market variations? How does the market solve the problem of increasing the availability of timber in periods of peak demand and storing potentially harvestable timber on the "stump" during the troughs in the market?

Forest taxation is a policy issue that is closely associated with the question of timber supply. Different kinds of taxes represent policy alternatives and may be chosen to accomplish varying goals or ends. Some taxes may also affect social welfare by reducing timber supply and increasing stumpage prices and final forest products prices. Can there be an equitable forest tax under current insitutional constraints? How will forest taxes affect the supply of timber land? These problems all relate in one manner or another to the theory of timber supply. Given a positive theory of timber supply, answers to the above questions and others may be furnished.

Contemporary timber production often includes an integrated series of operations over the life of the stand community. The biological result of one operation is affected by previous operations and often takes into account expected future operations. The entire series of operations is referred to as a management regime and involves planning. Because timber supply is time dependent, determining the optimum regime is inseparable from investment analysis, planning, and the actual implementation of the plan. A good deal of the theory of the behavior of the private timber firm is hence a theory of planning.

THE NATURE OF THE RESEARCH PROBLEM

Any discussion of timber production necessarily involves capital theory or production over time. Timber originates from either a stock of seeds or vegatatively from the roots or stump of its parent. Trees gradually grow to a point where humans desire to fell them and convert them into some useable product or otherwise dispose of them. Understanding the nature of timber during its course to maturity is essential. From a biological point of view, the tree itself is a factory or form of capital. The leaves or needles combine energy, light, water, and nutrients into woody material. Production of wood involves measurable time and thus timber production is time dependent. The physical factors of production are the trees themselves and another composite set of factors which foresters call the "site." The site encompasses both stock and flow characteristics. Soil nutrients and physical characteristics such as the slope and aspect of the land are of a stock nature. Such inputs as sunlight and rainfall are of flow dimensions. The time rate of tree growth is essentially a function of the genetic characteristics of the tree, the role of the tree in its stand community, the site, and any other humanly determined manipulation.

In economics, time is not a factor of production. Rather, timing is the concern of the decision maker and will affect the value of the timber asset in the future as well as the present. By making the right decisions at the right time, the decision maker can alter both the physical composition of timber that occupies the site over time and its economic value.

Just as time is not a factor of production, time is costless. The rate of interest is not the cost of time but rather a rate of time preference. The rate of interest represents a premium for sooner as opposed to later availability of economic goods. In the purest sense, the rate of interest indicates a behavioral rate of time preference established in a market of borrowers and lenders (Fisher, 1930). These borrowers and lenders make their decisions in the context of both the marginal product of capital and their intertemporal consumption preferences.

The case of growing trees or aging wine in barrels has both fascinated and perplexed many economists. Trees or wine represent an asset, the value of which increases at least through some span of time. Trees are planted at some point in time and the economic question then is when to harvest them. In this elementary form, the problem is called a point-input-point-output problem. More complex cases entail several inputs and/or outputs.

Foresters and economists are indebted to Mason Gaffney (1957) for his important contribution which clarifies the theory of the optimum point of terminating a stand of trees in the point-input-point-output context. Gaffney indicates that where the objective function is to maximize wealth in an infinite series of plantings and harvests, the Faustmann solution is the theoretically correct method of determining the harvest age. Hirschleifer (1970) adds additional clarity to the question by indicating that subject to specific conditions, investment criteria which are consistent with the maximization of wealth lead to optimal investment choices. The theory of investment infers the maximization of net present value in investment decisions.

The point-input-point-output problem investigated by Gaffney employs a fixed value function over time. The value function is determined at each point in time by the per unit of volume stumpage price, less the per unit of volume removal costs, times the standing merchantable volume. In addition, Gaffney's model includes a fixed planting input and its associated financial outlay. Given the fixed planting outlay, the net asset value over time, and the interest rate, the harvest age which maximizes the owner's wealth is solved. There are two other pieces of information in the point-input-point-output model which have have market implications: the quantity produced at the time of harvest; and in the infinite series, the frequency of harvests.

RESEARCH OBJECTIVES

Given the introductory material on the nature of the economic problem along with the concern with the future availability of timber, the objectives of this book are more readily outlined. The reactions of most economists to the future timber supply controversy would likely be quite forthright. The normal theory of supply indicates that equilibrium output is a positive function of product price. The nature of the theory of supply indicates that as the price of timber increases, more timber will be available for harvest. However, in developing the conventional theory of supply, economists resort to a world of static production. To date, no theory of timber supply has been developed in which production itself requires time.

As a result, the major objective of this study is to develop a theory of timber supply under conditions in which production is time dependent. In order to come to grips with the theory of timber supply, the analysis borrows heavily from earlier modes of investigation. For example, the basic microeconomic unit is the firm. The theory of the private competitive timber firm is given fairly broad analysis in order to determine how optimum investment and output respond to changing market stimuli.

The second stage of the analysis is the industry or market itself. It has long been recognized that the process of inter-firm competition for scarce productive factors itself influences market supply. In aggregating the member firms into an industry supply function, such considerations as the impact of different industry output levels on the costs of productive factors are taken into account. The market level of analysis is also the vehicle used to indicate output changes in response to changing demand conditions.

Once the theory of the private timber market is developed, the model will be used to study the incidence of forest taxes. The theory of timber supply is a necessary step in investigating the forest tax issue. For example, in order to determine whether or not any forest tax is neutral, one must determine whether the tax shifts industry supply. Another problem in the case of timber taxation is the mixed public nature of the industry. The taxation study is conducted under circumstances in which taxes are levied only on the private sector in order to indicate changes in intersectoral competitive advantage.

The theoretical model developed will be used to obtain valuable inferences concerning the economics of timber production planning in light of opportunity costs. These inferences appear to be original and add an important perspective to the theory of the firm in general. Other findings regarding the effect of changes in interest rates on equilibrium output and prices bear some importance in a theoretical context as well. However, the development of a theoretically stable market structure for timber under private land tenure and the findings regarding forest taxation are the highlights of the study.

RELATIONSHIP OF THE STUDY TO PREVIOUS RESEARCH

This study is an extension of Gaffney's (1957) influential "Concepts of the Financial Maturity of Timber and Other Assets." The attention of that study was the critical examination of varying criteria for determining the optimum maturity or duration of growth for timber or other appreciating assets. In Gaffney's study, the management regime was held constant and consisted of a planting input and its associated financial outlay on some homogeneous and unchanging forest site. Typically, other decisions in addition to the optimum time length of growth are made in the course of producing timber. Decisions, for example, regarding the choice of species, whether to thin the stand, pruning, or fertilization are often considered. Each of these decisions must be made taking into account the others in order to derive an optimum management plan.

Gaffney's work is properly seen as a portion of the theory of the private timber firm. It is implicit in his approach that the firm is a price taker or private member of a competitive industry. The monopoly theory of timber production has also been investigated (Walker, 1971). In this later study, investment and financial maturity were determined so that the firm could control industry output at a point where intertemporal marginal cost and marginal revenue were equated. To date, no one has studied the timber industry under conditions of competition, private land tenure and a time dependent production possibilities function.

Because most social welfare theorems are based on competitive conditions in production, the theory of competitive timber production and supply provides a vital linkage for the analysis of public forest policies. Again, however, welfare propositions are normally formulated under static conditions of production and consumption (Bator, 1957). The particular welfare proposition focused on in this study is efficiency in production, a necessary condition for the attainment of an optimum welfare frontier. As a result, the welfare criterion used is the payment of any surpluses beyond those which are required to employ a particular factor in production (Lerner, 1944, pp. 212-227). The welfare implications of what is often referred to as "soil rent" are alluded to in the section on forest taxes and the final chapter covering research implications.

RESEARCH APPROACH AND METHODOLOGY

The study starts at the traditional point of departure in economic theory, the individual operating unit or firm. Most of the analysis uses a simplistic view of the production alternatives facing the firm. As the analysis is developed, several of the simplifying assumptions will be relaxed. The use of simplifying assumptions is justifiable on grounds that theoretical considerations are more readily apparent.

The major portion of the analysis involves two decision variables for the firm. The two variable case is investigated under varying assumptions about changing timber quality and stumpage prices over time. Following the two variable analysis in the theory of the firm, an outline of the theory of production with a multiple-point-input-multiple-point-output model is presented.

The actual methodology used is that of comparative statics (Samuelson, 1947). Use of comparative statics may be misleading in the light of time dependent production. The meaning of comparative statics analysis is the displacement of some initial stationary optimum combination of production decisions by an exogeneous shift in a parameter of the model and the direction of movement of the variables included in the model to the subsequent stationary equilibrium state. It should be made clear that corner solutions in the model are ignored by assumption. Discussion of the n-decision alternative production model is more in accord with production decisions in the light of corner solutions.

A question may arise as to whether the purpose of the research is to indicate how to operationally manage timber or to postulate the general framework as to how private producers operate. Obviously, the two questions are closely related. Predictive theories usually rely on the assumption of continuous functions, while operational decisions are made in the light of discontinuous or discrete production alternatives. In attempting to postulate predictive theories as to how timber firms behave, (the case of this study) many of the operational complexities are suppressed. The firm is treated in a very mechanistic manner as if it were some organism responding to exogeneous market stimuli. In some instances, however, the analytical results contain overwhelming managerial significance. In these cases, the reader's attention is explicitly drawn to those points.

GENERAL ASSUMPTIONS AND DEFINITIONS

The phrase "the firm" has been loosely referred to without any attempt to more carefully define its meaning. In this study, the firm is seen largely as an entity which both holds the rights to use productive factors and organizes the factors to produce timber. The firm thereby in part replaces the market mechanism in the organization of production. For reasons of efficiency, inter-firm allocation decisions are non market in nature (Coase, 1937). The main distinction between this analysis and the more conventional theory of the firm (see for example Henderson and Quandt, 1958), is that the revenue associated with the employment of factors of production is commonly expended at different dates from the timing of receipts resulting from product sales. While these issues have been studied to some considerable extent (Hirschleifer, 1970; Gaffney, 1957, F. Lutz and V. Lutz, 1957), the integration of marginal investment and production decisions on the part of the timber firm in a competitive context is unique.

The Economic Environment

The objective function for the typical firm is assumed to be the maximization of wealth attendant with the ownership of its land and purchase of variable factors of production. In order for the firm to exist in a state of equilibrium, the production of timber must be the highest valued alternative use of its land. We assume that all potential timber land has some positive value in use, so that marginal land is that land for which the owner is indifferent between employing it in timber or selling it for some other productive use (see Appendix A.1). Thus, the firm's criterion for producing timber is whether the production of timber is the decision which will maximize the value of the use of its land. Marginal land may have higher or lower biological productivity than non marginal land. Finally, timber is the only valued product attendant with the land and there are no other non specific benefits or costs.

The firm in this analysis is a competitive member of the timber industry. The industry has both private and public sectors. The major emphasis is on the private sector with only brief attention given to the subject matter of public investment and supply. The competitive firm, being a price taker, exerts no noticeable leverage on product prices, factor prices, and the capital market.

The typical firm is not vertically integrated in the larger forest products industry. It owns no mill capacity nor wholesale or retail outlets. This assumption avoids what Duerr (1960) calls type-C costs, the cost of maintaining some degree of continuity of mill supply over time.

In the initial analysis, one final assumption is warranted. The net stumpage value of a stand is simply the gross value per unit of volume less the cost of harvesting and hauling and is independent of the total volume or area harvested.

The Biotic and Edaphic Milieu

In order to further simplify the analysis, it is assumed that the land owned by the firm is homogeneous with respect to productive capacity. Any new land entering timber production will be seen as an entering new firm(s) and the reduction of land is therefore an exiting of firms. Constant returns to scale with respect to land exist throughout the industry. No firm in the industry has any old growth timber in its inventory. The study is therefore characterized by tree farming and ignores the question of the rate of reduction of old growth timber supplies. The merchantable species is an intolerant pioneer so that even-aged management and clear cutting are the only assumed technological alternatives to successive reproductions on the same unit of land.

Market and Product Information

Statistical variation in forest production is an everyday fact of life. Fluctuations in weather patterns and stand damage due to fire and pests are perhaps the most common agents underlying production variability. As important as are variations in production, are the changes in market conditions which affect both production costs and product values. It is assumed that all density functions are normally distributed so that the mean of the density function is the best estimate of the occurrence of the event. We further assume that the producers have perfect information of these density functions.

ORGANIZATION OF THE STUDY

The following chapter contains the theory of the private firm. The theory of the firm in turn contains several subsections which are differentiated by varying assumptions about production opportunities and market conditions. Most of the analysis is concerned with the production of a homogeneous timber product but that assumption is relaxed in the section covering production with heterogeneous products.

The third chapter considers the problem of aggregated production and output in the market. One section in this chapter presents a brief review of output in the public sector under the policy constraints of sustained yield.

Chapter IV considers the incidence of forest taxes in a mixed public-private timber industry. The yield tax, income tax, and property tax are studied in the context of the market in order to determine their impact on economic efficiency, output and price. The results of the taxation analysis are contrasted with previous theoretical studies of forest taxation.

The final chapter draws together the policy implications and suggestions for further research. Attention is given in this chapter to limitations of the research, welfare considerations, and also to administrative problems of the various forest taxes.

2 The Theory of the Private Timber Firm

INTRODUCTION

The main purposes of this chapter are twofold. First, to present an analysis of those factors considered to be the most pertinent in determining the behavior of a private timber firm. Much of the material presented in this chapter will be integrated into the consideration of the behavior of several private firms in the market in subsequent chapters. Second, an initial two-decision variable production opportunity set is defined for the firm in order to clarify many of the issues of the effects of changes in market determined parameters. The two variable model will be used in the subsequent market aggregation of producers.

Following the two variable models, the theory of production with n-point-inputs and a single point-output is presented. This in turn gives way to the study of production with an n-point-input m-point-output analysis. In these latter instances, propositions are developed for the necessary conditions of an optimum management regime.

The subject of speculation both in terms of the market and biological variability in growth is the final area of analysis. This section illustrates how the firm adapts its management plan to short term changes in market conditions as well as to deviations from the expected timber yield.

NATURE OF THE PRODUCTION FUNCTION

In introducing the subject matter of timber production, two decision variables will be utilized. The duration of growth or life of the stand is designated as (t) years, and an establishment input at the point of origination of a new stand is (X) units. In reality, establishment of a new stand involves several inputs and alternative methods of regeneration. However, this analysis treats the establishment operation as a homogeneous physical input (X), because it helps to separate the induced changes in the management plan resulting from changes in establishment costs as opposed to other market changes. The forestry literature generally sanctions the term "management intensity." In a sense, the variable, (X), could be easily viewed as the "intensity of management." Particularly in the two variable production functions, readers could substitute the phrases "intensity of management" for "establishment" without altering the meaning of the results. In more complex production functions, it is often difficult to identify which alternatives are of the greatest management intensity.

The particularly relevant lower bound of the establishment input is zero, or natural regeneration. The fixed factor, the site, includes the entire set of ecological inputs which either positively or negatively influence stand growth and volume overtime. The harvest of the stand effectively determines the period of time for which the productive services encompassed in the site are available to the stand community. The establishment input influences the total volume in the stand community supported by the site at any point in time. Furthermore, the site is available continuously, while the productivity of the establishment input is terminated at harvest time.

Figure 2.1 indicates the production function for various combinations of (X) and (t). V_n is the lower bound of the set and indicates expected volume per acre over time for natually regenerated stands on the homogeneous site. As the level of (X) employed increased, the yield function approaches V_f, the volume per acre over time for a so called "fully stocked" stand. We shall ignore the occasional problem of over stocking and indicate the area bounded by V_n and V_f as the continuously differentiable production function set.

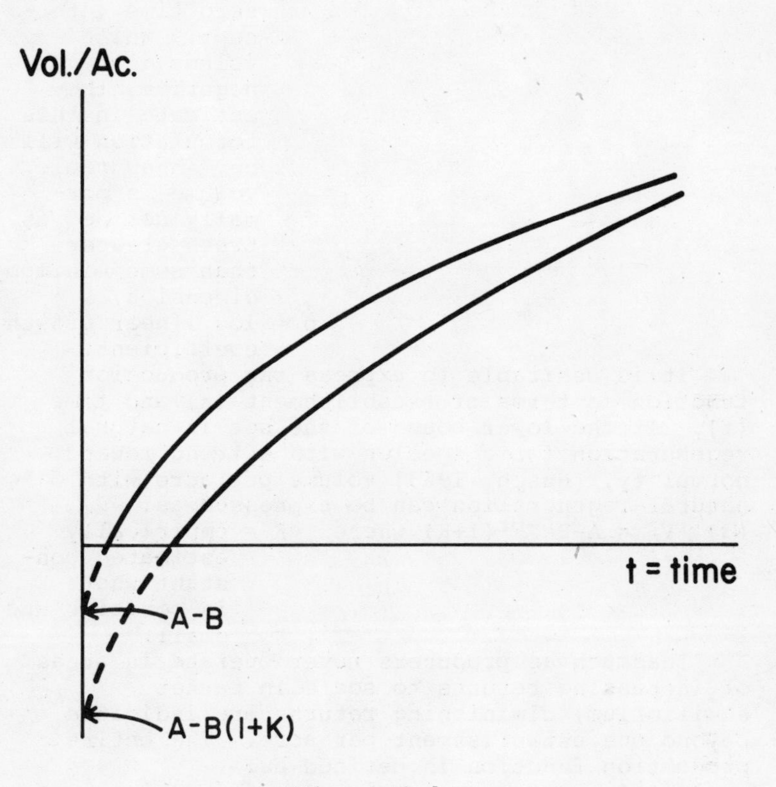

Figure 2.1

(1) $V = f(X,t)$ where: V = Volume per acre
X= No. of physical establishment units employed per acre.
t= time length of growth.

The volume per acre for a fully stocked stand shall be expressed with a natural growth function.

(2) $V_F = A-Be^{-bt}$ where. A = a theoretical maximum volume or upper bound of the biological capacity of the site per unit per acre.

A-B = the estimated zero time intercept. While volume is never negative, the estimate in this formulation will be. Measurable volume is normally defined as trees greater than some minimum dimension.

b = log linear growth coefficient.

It is desirable to express the production function in terms of establishment (X) and time (t). If the lower bound of the set is natural regeneration for a species with a trend toward normality, (Husch, 1963) volume per acre with natural regeneration can be expressed as:

(3) $V_n = A\text{-}Be^{-bt}(1+K)$ where K = empirically estimated constant where K is positive and small.

Inasmuch as producers never operate in areas of increasing returns to scale in market equilibrium, diminishing returns are indicated beyond one establishment per acre. The entire production function is defined as:

(4) $V = f(x,t) = A\text{-}Be^{-bt}(1+Ke^{-X})$ where $1 < X < \infty$

For values of (X) greater than one, volume per acre increases at a decreasing rate at all points in time as (X) approaches positive infinity. By limiting the value of (X) between one and infinity, we can ignore corner solutions and increasing returns to scale. The fact that volume per acre is never expected to follow the fully stocked yield path has real world relevance in light of the previously mentioned risks in production such as forest fires.

The production function indicated in Figure 2.1 is defined by equation (4). Expressing the production function in such a manner offers several analytical luxuries. The function f(X,t) is continuously differentiable with respect to the two decision variables and it reflects to a great degree the manner by which planting or establishment decisions are expected to affect volume in the range of ages relevant to maturity. In actual investment analysis, the physical production resulting from different input levels is characterized by lumpiness or several discrete alternatives. The formulation here is consistent with the belief that there are diminishing returns to the level of invested funds in establishment.

THE TIMBER INVESTMENT PLAN

By assumption in Chapter I, it was indicated that the value of the land in timber production is positive. The infinite series of harvests is therefore the correct analytical concept for determining harvest age. Appendix A.1 suggests the role that market determined parameters such as factor costs, product prices, or the discount rate, play in determining the choice between the infinite series and single series rotation. The objective function for the firm is summarized in equation (5).

$$(5)\quad W = WXe^{-0it} + (PV-wX)e^{1it} + (PV-wX)e^{-2it} + \ldots + (PV-wX)e^{-nit}$$

mult. by $1 = (e^{it}-1)/(e^{it}-1)$

$$W = \frac{PV-wXe^{it}}{e^{it}-1}$$

MAXIMIZE $W = \dfrac{PV-wXe^{it}}{e^{it}-1}$ where

P = net stumpage price. Assumed here to be constant per unit of volume and independent of number of acres harvested, and and constant over time.

W = wealth

V = vol. per acre harvested, a function of (X) and (t)

i = market determined real

opportunity cost of invested funds
X = no. of establishment units employed per acre
w = per unit cost of establishment
t = time length of growth

Wealth is a function of two decision variables, and three market determined parameters (Samuelson, 1947, p. 10).

$$(6) \quad W = g(X,t; P,i,w)$$

In formulating the long run management plan, the optimal investment in (X) will depend upon the estimated time of maturity (t) and vice versa. Each of the two decision variables must be solved taking into account the extent of the other. Because the firm can borrow (unlimited) funds at the rate i, the problem of determining the optimum X^* and t_m is solved by the unconstrained maximization technique. The partial derivatives of the objective function with respect to the two decision variables are set equal to zero and solved simultaneously for X and t. The first order conditions for maximization are:

$$(5a) \quad P(\frac{\partial V}{\partial X}) - we^{it} = 0$$

$$(5b) \quad P(\frac{\partial V}{\partial t}) = \frac{i(PV-wX)}{1 - e^{it}} = P(\frac{\partial V}{\partial t}) - i\frac{PV-wXe^{it}}{e^{it} - 1} + iPV = 0$$

(See Appendix A.1)

Equation (5a) indicates that holding the time of maturity constant, the firm will invest in establishment inputs until, at the margin, the value added by the last unit of input is equal to its compounded cost. Equation (5b) is the Faustmann (1849) condition. Holding (X) constant, the firm plans to grow the stand until the intertemporal change in value is equal to the rent (a flow = $i(PV-wXe^{it})/(e^{it}-1))$ plus the opportunity cost of the timber itself (iPV). We note that rent is a return to the employment of both the fixed factor (the site) and the variable factor (X). Equation (5a) is a marginal solution. Given diminishing returns, the total revenue product associated with X* is greater than the total cost.[1] With variable factors of production in addition to the fixed site factor, the phrases, "soil rent", "land rent", or "soil expectation value", may be misleading. Varying X changes the marginal product, average product and return to the fixed factor (the site). The rent indicated in equation (5b) is a maximum only with the optimum combination of both (X) and (t).

Equations (5a) and (5b) represent the necessary conditions for a maximum. It is important to check the second order or sufficient conditions to insure that we have not found either a saddle point or minimum solution. The second order direct partial derivatives must be negative and the second order partial derivatives of the objective function form a 2x2 Hessian determinant that must be positive for the solution to be a maximum (Allen, 1938, pp. 422-520).

As Appendix A.2 indicates, the determinant is positive definite if $(b - iKe^{-X})$ is positive. The growth coefficient (b) must be greater than the interest rate weighted by the establishment level, a very realistic result. There is one other important attribute indicated by the Hessian determinant in Appendix A.2. Element $a_{12} = a_{21}$, formed by the second order cross-partial derivatives of the objective function, is negative. If time and establishment were both factors of production in terms of physical inputs, the sign would indicate tht they were substitute factors of production (Allen, 1938). The economic nature of time makes the interpretation of this sign more difficult. As time increases, holding the stocking input constant and the site constant, the marginal product of planting or establishment falls. While this point may appear fairly esoteric, its importance will reveal itself shortly. Perhaps more clearly, the decision variables of optimum rotation age and stocking are substitutes.

The meaning of time and establishment as decision substitutes can be seen by referring back to Figure 2.1. At any particular positive point in time, an increase in (X) increases volume within the production opportunity set. However, holding that particular point in time constant, the slope or time rate of growth decreases. Hence, holding time constant an increase in (X) reduces the rate of change in output associated with time. Later in this chapter, production is considered in a different set of assumptions concerning growth and in that section, time complements the productivity of establishment.

COMPARATIVE STATISTICS RESULTS

We can now determine the direction of movement in the decision variables resulting from an exogeneous shift in one of the parameters (i,P, or w). By limiting the values of X between greater than one and less than infinity via assumption, the analysis is only slightly forced. All changes in parameters will produce changes in the management plan. In reality, a range of prices may warrant natural regeneration (X = 0). Comparative statics analysis is indicative of small changes in parameters so that previously unemployed production alternatives are not included in the revised management plan. The actual analysis is presented in Appendices A.3 -A.5.

Perhaps of utmost importance is the effect of changes in net stumpage prices on the real physical level of investment, in (X), and the optimum rotation age (t). As Appendix A.3 indicates, once and for all shifts in stumpage prices imply that:

$$(7)\ \frac{\partial X}{\partial P} > 0;\ \frac{\partial t}{\partial P} < 0.$$

If the price increases, the management plan will call for a higher physical establishment investment and shorter rotation age. The long run effect would be the substitution of establishment for the time length of growth. This results from our assumption of the trend toward normality, time and stocking are decision substitutes.

The effect of a change in the cost of establishment (w) has been a source of confusion since Gaffney's publication of "Concepts of the Financial Maturity of Timber and Other Assets." His analysis of a fixed yield function included the planting input as a fixed factor of production in addition to the site. The only decision variable was the time of maturity. Gaffney correctly indicated that the rotation age was a positive function of planting cost. Gaffney found that the effect of an increase in the planting cost (holding the physical planting input constant), was to increase the rotation age. He also correctly observed that the rotation age for the single rotation was longer than the infinite series rotation age. Hence, the effect of a higher planting cost was to move the stand toward the point of marginality where the land would be allocated to some other productive use (single series). Gaffney's analysis is consistent with this one, holding equation (5a) constant. Defining the length of the run by the number of variables held constant (i.e., long run versus short run), Gaffney's analysis is at best a short run solution. It is indicated later that opportunity costs change instantaneously while implementation of optimum plans takes time. In this model, the effect of a change in the factor cost of establishment (w) on optimum maturity and optimum establishment is:

(8) $\frac{\partial X}{\partial w} < 0; \frac{\partial t}{\partial w} > 0$ (See Appendix A.4)

Where time and the establishment decision are substitutes, a higher establishment cost implies that the rational producer will substitute more time for the establishment investment.

The last parameter included in the model is the market rate of interest. It is commonly believed that the effect of an increase in the interest rate is to shorten the optimal duration in the infinite series of rotations (see Gaffney, 1957). Again, the results here differ from the conventional wisdom in the long run. As is proven in Appendix A.5, the effect of an increase in (i) is to <u>lengthen</u> the rotation age and lower the number of stocking inputs in the long run management plan.

(9) $\frac{\partial X}{\partial i} < 0; \frac{\partial t}{\partial i} > 0$

Where establishment and the time length of growth are decision substitutes in the production of timber, the rational producer would choose a lower yield function and allow the stand to grow for a longer period of time with a parametric increase in (i). Had establishment been held constant, an increase in (i) would result in an earlier optimum rotation age. In the event that the yield function is fixed and unalterable, an increase in (i) would result in a faster liquidation of stands. In the model presented here, the opposite is true in the long run. The operator substitutes more time for a lower expected yield function.

THE FIRM'S SUPPLY FUNCTION

The discussion to this point has indicated the market determination of the firm's management plan. In this first approximation of the firm's supply function, it is assumed that the firm's production will be compared at two different steady state price levels in order to determine whether or not its output is a positive function of price in the long run. The development of a firm's supply behavior is a necessary step in the subsequent determination of stability in the market context.

The production function, along with the market parameters, determine the expected output produced at a planned date with the optimal input level of the variable factor (X) of production. Output enters the market every (t_m) years. The flow per acre per unit of time under an initial optimum management plan is V^*/t_m for the optimal X^* where V^* is the planned volume at maturity. In the forestry literature, volume per acre harvested divided by the number of years taken to grow the timber is called mean annual increment (MAI). For a given level of (X), expected MAI is a function of (t). Hence, just as expected volume is a function of (X and t), so is expected MAI, or flow per acre, a function of (X and t).

The supply question, therefore, can be phrased in terms familiar to the forester (Vaux, 1973). What is the effect of a change in price on the optimum MAI, or flow per acre? There are two aspects of the supply question: how much volume is available at maturity; and the frequency of harvests (maturities) for each particular acre of land in production. In other words, the supply problem is of both frequency and quantity dimensions. In the conventional theory of supply with a static production function, changing supply is only of a quantity dimension per unit of time. It was indicated that the optimum maturity is a negative function of the product price ($\partial t/\partial P < 0$). Thus, the <u>frequency</u> of harvests <u>increases</u> in the infinite series of harvests with a higher price. It was also proven that the number of establishment inputs employed was an increasing function of the price ($\partial X/\partial P > 0$).

As is indicated in Figure 2.2 below, with an initial combination at point C on $V = f(X^*, t_m)$ at volume V^* and optimum rotation t_m, an increase in P will move the optimum combination to a higher yield function such as $V = f(X^{**}, t_m)$.

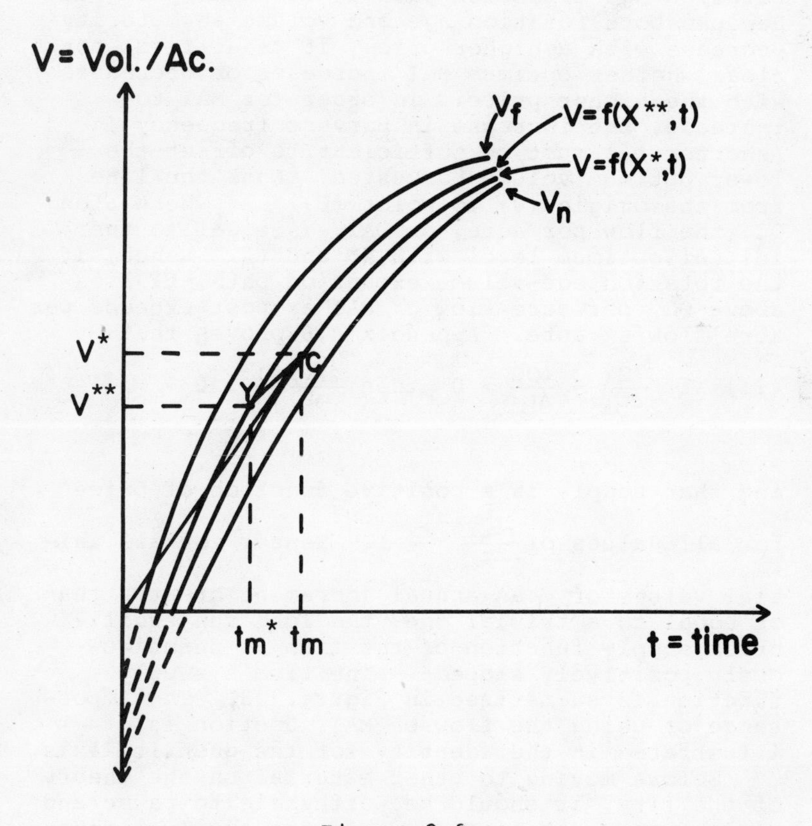

Figure 2.2

By the implicit function theorem, we find that the new planned volume at the shorter rotation age (t_m) is a negative function of the price

$$(10)\ \frac{\partial V}{\partial P} = \frac{\partial V}{\partial X}\frac{\partial X}{\partial P} + \frac{\partial V}{\partial t}\frac{\partial t}{\partial P} < 0 \qquad \text{(Appendix A.6)}$$

$$\text{Note: } \frac{\partial V}{\partial X}\frac{\partial X}{\partial P} > 0 \text{ and } \frac{\partial V}{\partial t}\frac{\partial t}{\partial P} < 0$$

Thus, as is indicated in Figure 2.2, the new optimum combination under the higher price yields both a lower volume ($V^{**} < V^*$) and a shorter rotation age ($t_m^* < t_m$). We shall define the volume rotation age expansion path as CY in the diagram. Because both rotation age and volume at maturity decrease with a higher price, it is at first not clear whether optimum MAI increases or decreases with the higher price. In order for MAI to increase, the increase in harvest frequency (shorter t_m) must be sufficient to offset the lower optimum volume harvested. Note the line from the origin (0) to point (C). Anywhere along OC, the flow per acre, or MAI, is equal to the initial optimum level with X^* and t_m. Hence, if the rotation age-volume expansion path, CY lies above OC, per acre flow or MAI ex post exceeds per acre flow ex ante. Appendix A.6 proves that:

$$(11) \text{ if } \frac{\Delta CY}{\Delta t_m} - \frac{\Delta OC}{\Delta t_m} < 0 \text{ then } \frac{\partial (V/t)}{\partial P} > 0$$

and that supply is a positive function of price for all values of $\frac{\Delta CY}{\Delta t} \geq 1$. Hence, for all initial values of mean annual increment greater than or equal to a trivial one, the long run equilibrium supply function of the firm is unambiguously positively sloped[2]. The firm's supply function is summarized in Figure 2.3. The importance of using the flow or MAI function is illustrated in the identity for the quantity axis.

Before moving to other material on the theory of the firm, it should be worthwhile to pause and reflect on these findings. If the supply function for the firm had been ambiguous or negatively sloped, the market implications to be developed in a subsequent chapter would have been dramatic indeed. Holding the supply of land constant, the equilibrium private sector supply could have been negative giving vent to the possibility of instability. The findings presented here rule out the problem of instability and any potential policy questions that might have been implied.

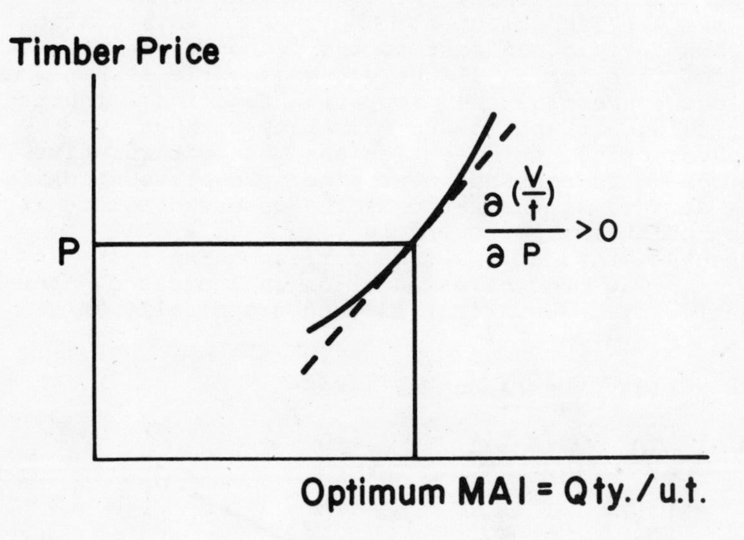

Figure 2.3

In the previous section, it was indicated that the long run effect of a change (increase) in (i) was to reduce the establishment level and lengthen rotation age. However, the more important implications lie with the effect of change in (i) on the firm's supply function. As is indicated in Appendix A.7, volume at maturity is inversely related to the interest rate.

$$(12) \frac{\partial V}{\partial i} = (\partial V/\partial i)(\partial X/\partial i) + (\partial V/\partial t)(\partial t/\partial i) < 0$$

Because a longer optimum rotation age implies a reduction in the frequency of harvests and, as indicated above, lower volume at the time of harvests, an increase in (i) results in a reduction (shift) in the firm's supply function for the same price level and factor cost of establishment.

ESTABLISHMENT AND ROTATION AGE:
DECISION COMPLEMENTS

The findings thus far have been based on a production function similar to the empirical relationship of growth with a trend to normality without overstocking (Husch, 1963). The result was that the establishment and rotation age decisions were substitutes (Eqs. 7-9). This section draws a vivid contrast to the propositions developed thus far by slightly changing the assumptions concerning the production function. Instead of using a trend toward normality without overstocking, this section assumes that relative stocking is constant over time. Relative stocking is defined as the ratio of the observed volume at any point in time over the volume of a fully stocked stand at the same corresponding point in time. The production function is indicated mathematically in equation (13) and graphically in Figure 2.4.

$$(13) \quad V = (A - Be^{-bt})(1 - Ke^{-X})$$

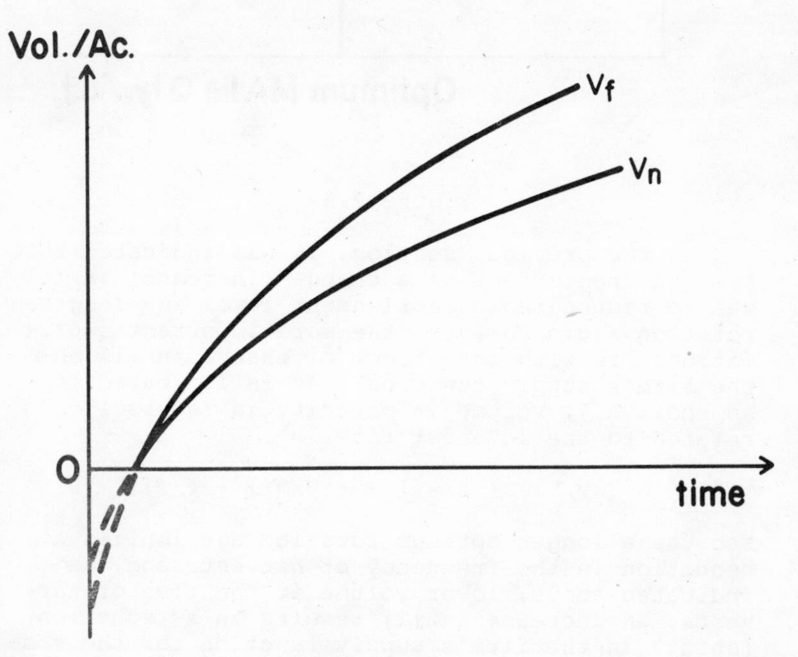

Figure 2.4

As Appendix A.8 indicates, the convexity of the function implies the necessary conditions are easily satisfied for a maximization, but the sign of $b_{12} = b_{21}$ is positive. The decision variables: stocking and time, are complementary in production.[3]

At any particular positive point in time, an increase in (X) increases the slope of the associated volume function within the opportunity set. Increased establishment increases the time rate of change of volume holding time constant. Under this changed production assumption, the effect of a change in price is different from the analysis presented under the earlier assumption.

$$(14) \quad \frac{\partial t}{\partial P} > 0; \quad \frac{\partial X}{\partial P} > 0 \qquad \text{(See Appendix A.8)}$$

The result should be heavily underscored. There is no economic reason to believe that the long run rotation age will necessarily shorten with higher prices. It depends upon whether the length of maturity is a net complement or substitute for other decision variables. If the empirical observation that higher stumpage prices in general result in shorter rotation ages is true (as it appears to be), it is the result of the fact that the duration of growth is a net substitute for other decision variables.

CHANGES IN REAL PRICES AND MANAGEMENT OVER TIME

The limits of rationality can be thought of as falling between two extremes; the perfectly rational pool player versus the myopic dog. The rational pool player knows all of the classic laws of physics and fully understands that each shot is part of a strategic series of events. On the other hand, the myopic dog when responding to the sound of his walking master's whistle, solves the problem by running in a curve. Unable to predict his master's walking speed and his own appropriate running speed and vector path required to solve the problem most efficiently, the dog expends excess energy.

Whether any adjustment process is more like the myopic dog or the rational pool player is a matter of observation. What is presented here is the myopic dog solution to the determination of the optimum investment and rotation in the light of a constant time rate of change in one of the market parameters; the market price of stumpage.

Observing historical time series data of stumpage prices suggests that they behave somewhat spasmodically. In this section, a smooth trend of price changes or expected real stumpage prices, $E(\dot{P}/P)$ is assumed and ia included as a parameter. The entrepreneur takes into account the time rate of change of prices but solves each management regime one at a time holding establishment and duration in the subsequent rotations equal to the initial one being solved. The perfectly rational approach to the problem would yield different rotation lengths and investment levels in each subsequent rotation.[4]

In the ensuing analysis, we shall see that once and for all changes in prices yield the same qualitative movement in the management variables as do gradual unidirectional changes in prices.

Consider Figure 2.5 below. Real stumpage prices are changing at a constant percentage rate relative to the price of other goods. Even though the entrepreneur takes the rate of change in prices as constant, the base price at two different points in time will differ. Hence the effect of investigating a change in the base price is the same thing as developing a management plan at two different points in time.

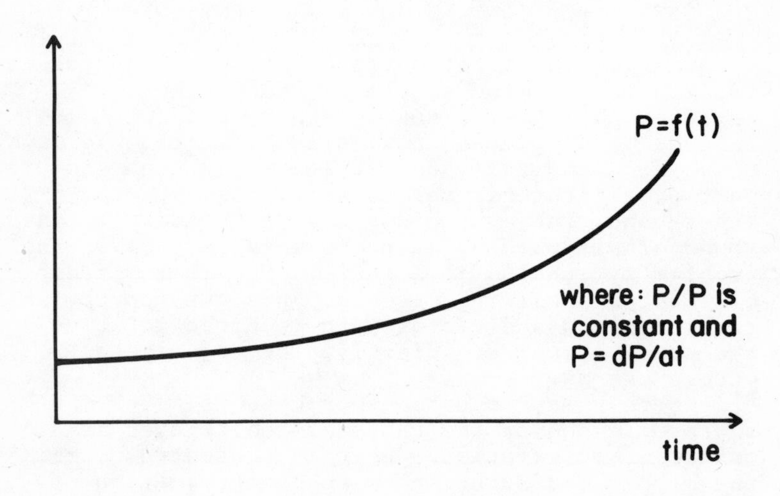

Figure 2.5

While the Minnesota Fats solution is not shown, the following intuitive results seem clear. First soil rent or bare land value will increase over time. This is based on the fact that revenues are increasing relative to costs. Second the most efficient solution for the current crop of trees takes into account the changing land value and the unique rotation and establishment investment for each unique subsequent crop in the infinite cycle of timber crops.

Again, we shall want to maximize wealth in the infinite series of rotations.

$$(15)\quad W' = h(X,t,;\ P,w,i,E(\dot{P}/P))$$

$$(16)\quad W' = P_{t_{m_1}} V(X,t_{m_1}) - wXe^{it_{m_1}} + \ldots$$

$$+ P_{t_{m_n}} V(S,t_{m_n}) - wXe^{it_{m_n}} =$$

$$\frac{PV}{e^{rt}-1} - \frac{wXe^{it}}{e^{it}-1}; \quad \text{where } r = i = i - E(\dot{P}/P) > 0$$

$$\dot{w}/w = 0$$

The first and second order conditions for maximization are indicated in Appendix A.9. We can then determine the effect of a change in price (P). The change in price can result from a once and for all change in demand at an instant in time, or from determining rotation at two different points in time along the curve in Figure 2.5 above. Thus, holding the rate of expected price increase constant, the factor cost of planting constant, and the real rate of interest constant, we find that:

$$(17)\quad \frac{\partial X}{\partial P} > 0; \quad \frac{\partial t}{\partial P} < 0 \qquad \text{(Appendix A.10)}$$

where time and stocking are substitutes. Had time and stocking been complementary, the results would have been consistent with equation (13).

It may be helpful to place these findings in a slightly broader context. The U.S. Forest Service, (1973, p. 148), has indicated an increasing long term historical trend in average net stumpage price of 2.5%. Assuming that the period of growth is a net substitute, as is commonly believed, one would predict that there has been a gradual shortening of the rotation age over time and a more intensive timber management. However, interpreting historical changes in management practices has some severe limitations. For example during the past 100 years, production technology has not been constant as has been assumed throughout this discussion. Changes in technology may have exacerbated or reduced the change in optimum rotation age that might have occurred under conditions of a constant state of the technological arts. For example, improved mechanization may have actually reduced the per acre cost of planting thereby exacerbating substitution.

Equally important in explaining changes in management practices over time is the fact that the methodological analysis is not invalid with a more realistic view of actual price changes. Once and for all changes in parameters give the same results as gradual changes in the parameters over time.

Somewhat lost in the theoretical discussion is a minor but rather important point of significance for investment analyses. It was explicitly assumed that the expected time rate of change of the factor input price was zero. (It was also implicitly assumed that the real rate of interest was constant over time.) Too frequently, investment analysts, apparently attempting to justify some silvicultural intervention, overlook the fact that real factor prices may be expected to change over time as well. We have, for example, recently seen dramatic changes in the cost of fertilizer. Whether or not these changes have been foreseeable does not preclude recognition of their possibility. As much attention must be paid to changing input price levels as to changing output price levels in order to arrive at sound investment decisions.

PRODUCTION WITH HETEROGENEOUS PRODUCT QUALITY

It is a well known fact that stands of mature and immature timber often consist of a wide variety of wood quality (McArdle, Meyer & Bruce, 1944). Variation of size, form, and internal cellular structure result in different per unit of volume values for stems of the same stand, species and age. As an even-aged stand of trees grows over time, individual members compete for growing space inclusive of light, water, and soil nutrients. Branches and roots extend, and the crown gradually closes, resulting in mortality of those members least fit to maintain their place in the stand community. (Spurr, 1964).

The production functions discussed previously indicate observable volume at different points in time and do not indicate the rate of mortality (Husch, 1963). As time passes, the available light, water, and nutrients tend to be concentrated in the dominant and codominant stand members, with the subordinate members tending toward, or entering mortality. In general, the codominant trees are of the best form with respect to their potentiality for sawlogs and peeler logs (plywood) and are, as a result, of the highest per unit of volume value. The dominant trees often possess flaws in their form which reduce their value, and the subordinate ones frequently exhibit larger degrees of rot or cost relatively more to handle and haul, which add to their relative inferiority. In addition, as the stand grows, natural pruning of lower branches which receive the least amount of sun occurs. Radial stem or trunk growth gradually covers previously branched areas and consists of knot free wood. In Douglas-fir, the outer area of the sawlog or peeler log is frequently referred to as "clear-fir" because it is knot free. Due to the structural characteristics of "clear-fir", the first sawlog (closest to the ground) is generally of higher per unit value than the rest of the tree.

This brief introduction of stand community development should indicate to non foresters some of the difficulties of stand appraisal. It will also serve to highlight some of the major reasons for the following discussion of stand value over time. The discussion may also, in part, serve to explain the existence of the complex system of grading harvested timber in order to facilitate market exchange.

Compared to the earlier analysis of the optimum management regime, the discussion of optimum stand management under conditions of heterogeneous product quality is somewhat less rigorous. Within the analysis in this section, however, the relationship between gross stumpage value, harvesting and hauling costs, and net stumpage value will be introduced.

Up to this point, the expected total revenue net of harvesting and hauling costs at the time of maturity has been indicated as:

(18) $NR = PV(X,t)$,

where (P) is the stumpage price net of hauling and harvesting costs. Maintaining consistency with the introduction to this section on the nature of heterogeneous products, the values of various log qualities can now be introduced.

Let Pg_j = the gross stumpage price (per unit of volume) of the jth log grade delivered to some destination point of n grades;
V_j = volume/acre of the jth grade;
C_1 = cost of harvesting inclusive of road construction costs = $a + bV$ (linear by assumption):
C_2 = per unit cost of hauling
C_H = $C_1 + C_2$; the cost of hauling and harvesting per unit volume.

Hence, in keeping with the general formulation presented, the total net revenue is:

(19) $$NR = \sum_{j=1}^{n} Pg_j V_j(X,t) - a + (b+C_2) \sum_{j=1}^{n} V_j \quad *$$

*Subject to the condition that the per gross unit stumpage value of the lowest quality log is greater than C_2 where only the volume which pays its way in hauling is considered.

Empirical work by Jackson & McQuillan (forthcoming) confirms this kind of relationship. In this case, timber value is a function of tree diameter. The value or net stumpage per unit of volume increases at a decreasing rate. For small diameter trees, the value is negative indicating that harvesting and hauling cost exceed delivered log values.

Holding relative real gross stumpage prices over time, establishment costs and harvesting costs constant, the intertemporal change in net revenue is:

(20) $$NR = (Pg_1 V_1 + \ldots Pg_n V_n) - (b+C_2) \sum_{j=1}^{n} V_j$$

Specification of stand values over time should utilize a marginal valuation context. All too frequently, average stumpage price is used as a proxy for the marginal concept. While the number of log grades and their corresponding relative values over time makes the analysis more complex, use of average stumpage price may distort the calculated rotation from the optimum value by a considerable amount. If the intertemporal marginal revenue function is estimated using average stumpages and the actual change in value over time is greater than that estimated, optimum maturity will be underestimated and vice versa.

One other extremely important point is indicated in this analysis. Equation (18) is the total net revenue subject to the condition that the per unit gross stumpage value of the lowest quality log was greater than (C_2) the per unit cost of hauling to the destination point. Because we are analyzing a clear-cutting environment, all trees on the site must fall. That does not necessarily mean that all felled material is commercially utilized. Thus, if the gross stumpage price of the lowest grade log (Pg_n) is less than the marginal cost of hauling (C_2), wealth maximizing behavior would result in leaving the total volume of the nth grade lying on the site as harvest residue (assuming residues don't affect future productivity).

Where production is characterized by heterogeneous products, the actual proportion of the standing biomass entering the market is determined by the relative prices and costs in a given technological state of product conversion. Indeed, one would predict that as the general real price level of wood products increases relative to production costs, utilization of standing biomass increases. Aggregate biomass utilized (supply) is a function of gross stumpage value, holding removal costs consatnat with heterogeneous product quality.

There has been a considerable discussion of total utilization of biomass (koch, 1974; Keays and Szabo, 1974; Keays and Hutton, 1974). Strict interpretation of total biomass utilization means use of roots, needles, bark and branches. Because a large portion of the stock of the site's nutrients are bound in some of these portions of the tree, it doesn't appear that the rational producer would be willing to make major depletions in his otherwise sustainable wealth via total utilization of biomass even if product conversion were feasible.

GENERALIZED PRODUCTION WITH n-DECISION ALTERNATIVES

The analysis to date has utilized two decision variables, (X and t). In order to maintain analytical simplicity, it is benefical to return to the assumption of homogeneous wood products and an expected time rate of change of zero for real factor and product prices. The more realistic problem than the two variable production function is the choice of the optimum set of alternatives from a larger set of technically possible ones. Not only must the producer choose the right alternatives, he must also determine the right time in the life of the stand to implement them. In this section, the underlying principles that determine an optimum fertilization regime without thinning will be presented and then the principles determining thinning will be discussed.

Figure 2.6 indicates the production function with a set of fertilizer inputs as well as the establishment input variable. Because the timing and intensity, or amount of fertilizer applied are continuous variables, the porduction space is bounded in a manner simmilar to the earlier two-variable production functions. Any point in an enclosed space is attainable with one or more combinations of inputs and input timings. The production function is:

(21) $V''=h(X,t,F_1,t_{f_1},\ldots,F_n,t_{f_n})$ where: X,t defined previously
F_j = fertilizer applied in lbs./ac. of jth input
t_{f_j} = stand age of jth input
$j = 1,\ldots,n$

In order to clarify the analysis, assume, as many biologists believe, that young stands respond more vigorously to fertilizer than old ones. (For an explanation of this assumption, see Kramer and Kozlowski, 1960, 168-169, 460-463). More precisely,

(21a) $\partial V''/\partial t_{f_j} < 0$ for all values of X,t,F

The revised objective function and attendant first order conditions appear below.

(22) Max. $$W''=\frac{PV''-fF_1e^{i(t-t_{f1})}-\ldots-fF_ne^{i(t-t_{fn})}-xXe^{it}}{e^{it}-1}$$

(22.1) $P(\frac{\partial V''}{\partial X})-we^{it}=0$

$\vdots$

(22.k+1) $P(\frac{\partial V''}{\partial F_j})-fe^{it}=0$

(22.k+2) $P(\frac{\partial V''}{\partial t_{f_j}})-ifF_je^{i(t-t_{f_j})}=0$

$\vdots$

(22.2n) $$P(\frac{\partial V''}{\partial t})-\frac{iPV''-wXe^{it}-fF_1e^{i(t_m-t_{f1})}-\ldots-fF_ne^{i(t-t_{fn})}}{e^{it}-1}$$

$-iPV''=0$

The above first order conditions should be carefully interpreted. Schreuder(1971), indica-

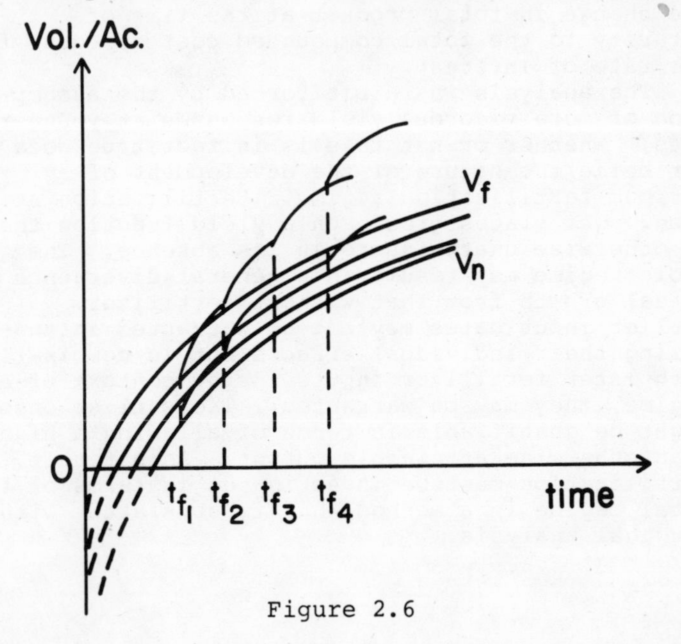

Figure 2.6

tes in an article on thinning that marginal analysis frequently involves the calculus of variations in order to arrive at optimum solutions. These necessary conditions are useful mainly in developing a theory of the optimal fertilization regime. Two conditions are of particular importance. Equation (21k+2) indicates the necessary condition for the optimum timing of the jth input. Referring to the assumption indicated in (20a), the condition for optimum timing for an intermediate input becomes obvious. Holding the other decision variables constant, the entrepreneur makes the jth input sufficiently early until the value of the marginal product associated with earlier timing equals the intertemporal marginal cost of the marginal change in the input date. In fact, we find that solving (21.j+2) for (i), we get a result which at first may appear surprising.

$$(23) \quad i = \frac{P \frac{\partial V''}{\partial t_{f_j}}}{fFe^{i(t-t_{F_j})}}$$

The entrepreneur will make the date of input sufficiently early so that the ratio of the value of the change in total product at the time of maturity to the total compounded cost is equal to the rate of interest.

The analysis was a bit forced by the assumption of more vigorous yield responses at younger ages. Whether or not this is in fact true does not belie the nature of the development of an optimum fertilization regime. Fertilization at young ages places growth on a yield function that is otherwise unattainable in its absence. The whole regime may result in a general divergence in actual growth from that without fertilizer. Earlier input dates may not be warranted in analyzing their individual effects but in combination with later fertilizer inputs in the context of a regime, they may be warranted. The earlier ones might be justifiable in terms of allowing a higher than otherwise attainable output. As a result, fertilization must be investigated in terms of the total regime in a method that is consistent with marginal analysis.

From an operational point of view, the theory defining the optimal limits of a regime should prove to be of considerable assistance. Specification of the regime as outlined above, indicates the conditions for the optimal frequency of inputs, intensity of each input, rotation age, and establishment. The assistance may likely take the form of constraints in programming problems based on the proper theoretical specification of the problems.

Turning now to another production alternative, thinning is commonly practiced in contemporary forestry. As Schreuder (1971) points out, there are several reasons for thinning stands. Since we are assuming choices involving a homogeneous product, one potential advantage of thinning is a change in the date of a portion of the revenue receipts. Furthermore, the operator may salvage mortality by reducing intra species competition.

The production function with a single thinning operation appears in Figure 2.7 and in functional form in equation (24).

$$V''' = k(X,T,t_T,t) \qquad (24)$$

where: T = % of total volume removed at time of thinning t_T

t_T = time of thinning

X,t = as before

While the yield function with thinning is discontinuous at time t_T, the function is still continuously differentiable because t_T and T are continuous decision variables. Once again, the objective function is revised to appear as:

$$W^* = \frac{P_g - C_T \; V(X,T,t_T)e^{i(t-t_T)} + P_g - C_H \; V(T,t_T t) - wXe^{it}}{e^{it}-1} \qquad (25)$$

where: $(P_g - C_T)$ = net unit value of thinning

$(P_g - C_H)$ = net per unit value of final harvest

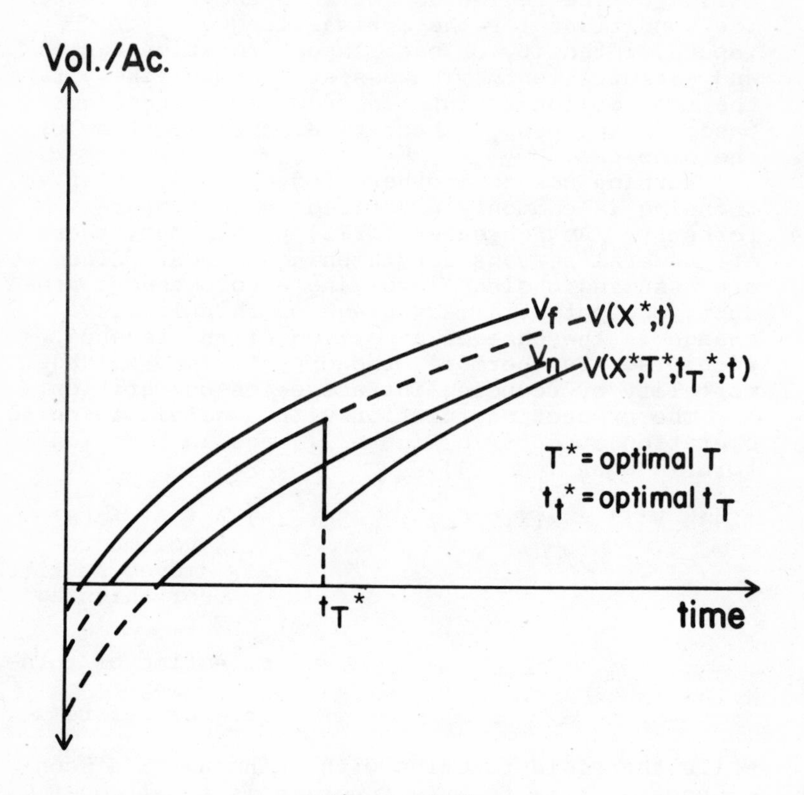

Figure 2.7

Looking at the necessary conditions for maximization, we find:

(25.1)

$$[P_g - C_T]\left(\frac{\partial V_{t_T}}{\partial X}\right) e^{it} + [P_g - C_H]\left(\frac{\partial V_{tm}}{\partial X}\right) - we^{it} = 0$$

(25.2)

$$[P_g - C_T]\left(\frac{\partial V_{t_T}}{\partial T}\right) e^{i(t-t_T)} + [P_g - C_H]\left(\frac{\partial V_{tm}}{\partial T}\right) = 0$$

(25.3)

$$[P_g - C_T]\, e^{i(t-t_T)}\left[\left(\frac{\partial V}{\partial t_T}\right) + iV^*_{t_T}\right] + (P_g - C_H)\left(\frac{\partial V_{tm}}{\partial t_T}\right) = 0$$

(25.4)

$$(P_g - C_H)\left(\frac{\partial V_{tm}}{\partial t}\right) -$$

$$i\left[\frac{(P_g - C_H)V_{t_T} e^{i(t-t_T)}}{e^{it}-1} + \frac{(P_g - C_H)V_t - wXe^{it}}{e^{it}-1} + (P_g - C_H)V_{t_1}\right] = 0$$

In this thinning example, condition (25.2) is of the greatest importance. The producer increases the percentage of material removed until the increase in value of the thinning is offset by the loss in value of the final harvest. While there may be increasing returns in lower levels of thinning, existence of positive interest rates imply that the producer will actually reduce the level of the final output in favor of the commercial thinning.

The principle of the optimal timing of the thinning operation is perhaps not quite so clear under conditions of homogeneous products. However, holding the percentage removed constant, earlier removal yields a lower total final output but a shorter period ($t-t_T$) to determine the present value. Thus, the optimum timing is determined when any further change in time will not increase total net present value of the combined thinning and final harvest values.

Where production is characterized by a heterogeneous product, a number of factors determine the optimal thinning regime, similar to (25.1-4) above. These factors include relative prices of the various log grades, and the various frequency distributions of associated stand compositions before and after thinning using different thinning techniques (e.g. low thinning, selection thinning, etc.; see Smith, 1962, pp. 64-124). Precommercial thinning, or thinning where the value of the product removed is less than the cost of the operation should be analyzed as an investment outlay.

The determination of the direction of change in rotation age with n-decision alternatives and discrete changes in stumpage price appears as a reasonable extension of the analysis. However, only a subjective appraisal is possible. Throughout certain ranges in prices, time may be a net substitute. This phenomenon would be the result of different optimum management regimes at different price levels inclusive of a different number of inputs. The only tentative conclusion possible about rotation age is that the optimum age may fluctuate as corner solutions are encountered and additional inputs added. If there is any generality in the movement accounting for the fluctuations for a range of prices, the direction of a movement can be attributed to the fact that time is either a net substitute or complement to the productivity of the set of inputs over time.

PRODUCTION AND MARKET SPECULATION

It may be surmised that the public forest management ideology of maximum sustained yield has led to a disproportionate emphasis in timber capital management on what shall be referred to as production speculation. The meaning of "production speculation" is the anticipation of growth and yield resulting from intervening natural and human inputs inclusive of the statistical variation of the outcomes.

On the other hand, market speculation is the anticipation of factor costs, interest rates and product price levels. In order to maintain continuity with the earlier sections, material concerning production speculation is presented first. the ordering should in no way infer a weighting of their relative importance. The underlying assumptions of the private firm imply an implicit ordering of all decision alternatives with respect to the impact on the firm's wealth.

To illustrate the economic nature of speculation on biological outcomes over time, a production function intermediate between the initial two factor case and the recent n-variable examples is utilized. Assume that the optimum long run production plan, (as was developed in Equations 21-24 above), includes the following decision variables: establishment (X), maturity (t), and a single fertilization input where time (t_f) and amount per acre (F) are variable. Utilizing a homogeneous product and holding the market determined parameters constant throughout the analysis, the objective function is:

$$(26) \qquad W^{**} = \frac{PV^{**} - wXe^{it} - fFe^{i(t-t_f)}}{e^{it}-1}$$

where: $V^{**} = A-Be^{-bt}(1+Ke^{-x}) + \dfrac{F^{\alpha_1} S_R^{\alpha_2} t_F^{-\alpha_3}}{[1+e^{-g(t-t_f)}]}$

$$S_R = \text{relative stocking} = \frac{A-Be^{-bt}(1+Ke^{-X})}{A-Be^{-bt}}$$

$(1+de^{-g(t-t_f)})$ = the fertilizer response over time;

$\alpha_1, \alpha_2, \alpha_3$ = all greater than 0 and less than 1; and other notation same as before

Again, the long run management plan is determined by the following necessary conditions:

$$(26.1) \quad P\; XKe^{-X}Be^{-bt} + \frac{F^{\alpha_1}\alpha_2 S_R^{\alpha_2-1} t_f^{-\alpha_3}}{1+de^{-g(t-t_f)}} - we^{it} = 0$$

$$(26.2) \quad P\; \frac{\alpha_1 F^{\alpha_1-1} S_R^{\alpha_2} t_f^{-\alpha_3}}{[1+de^{-g(t-t_f)}]} - fe^{i(t-t_f)} = 0$$

$$(26.3) \quad P[F^{\alpha_1} S_R^{\alpha_2} t_f^{-\alpha_3-1}[1+de^{-g(t-t_f)}] -gde^{-(t-t_f)} F^{\alpha_1} S_R^{\alpha_2} t_f^{-\alpha_3}] + ifFe^{i(t-t_f)} = 0$$

$$(26.4) \quad P\;(bBe^{-bt}(1+Ke^{-x}) + F^{\alpha_1} S_R^{\alpha_2} t_f^{-\alpha_3}\{\frac{1+de^{-g(t-t_f)} + gde^{-g(t-t_f)}}{(1+de^{-g(t-t_f)})^2}\}$$

(cont.) $$- i \frac{PV^{**} - fFe^{i(t-t_f)} - wXe^{it} + PV^{**}}{e^{it}-1} = 0$$

(26.5-26.n) (marginal value of all other alternatives less their associated marginal costs)

The above equations indicate long run speculation with all decisions variable. Now suppose the producer plants the anticipated optimal X* inputs but that the response varies from the original best estimate. Simply for reasons of a variation in weather conditions, mortality of a newly established stand may change. The producer must take the actual stocking into account in making his subsequent decisions. However, the "bare land value" or anticipated rent is still constant. The rent is the opportunity cost of removing the standing timber and reinitiating a new stand at any point in time. Hence, in taking account of actual stocking, the producer holds equation (26.1) constant, substitutes the revised production opportunity set into (26.2-n) and revises his management plan with the long term rent in equation (26.4) constant. Upon fertilization, if it is subsequently warranted, the producer can again survey the results and revise his production plans regarding any further inputs or the harvest date.

The distinguishing feature of production speculation is in the length of the run. The investment plan for regenerating timber on the site is the long run plan; as decisions are made and implemented, the length of the run for the current stand shortens. However, all decisions are made in the context of the long run opportunity cost of vacating the site. The short run decision for the current stand of timber is the final harvest decision, but once again the long run opportunity cost of the growing space is considered (holding all but (26.4) constant and substituting the actual yield function for the previously anticipated yield function). The gains and losses over time associated with the statistical variance in production take a form similar to windfall gains and losses. While they are not completely unaccounted for, they are variations from the best estimate of the flow of benefits and costs over time.

While market speculation is similar in form to production speculation, its distinctive character lies in the anticipation of movements in borrowing costs as well as product and factor prices. One should recognize again that there is perhaps an unparalleled planning horizon in timber production as compared to other contemporary non-forest activities. Use of the infinite series of rotations indicates that the producer expects a somewhat favorable market for his output forever. Expected revenues and costs are carried to infinity. Speculation may also take place in the intermediate run as well as the long run, but the market speculation indicated here will be of only the short run variety. It is important to note how the wealth maximizing producer reacts to short term variations in the stumpage price from the long term expected trend. The only decision variable in this analysis will be the actual harvest date.

Suppose, as is indicated in Figure 2.8, that time t_0 is the current point in time and that the dotted line represents the expected long term trend in gross stumpage prices. For simplicity, we indicate that the expected time rate of change in prices is zero, and that growth has behaved as anticipated. Holding all other decisions for the stand constant, the short run decision for maturity is:

$$(27) \quad (P_g - C_H)\left(\frac{\partial V}{\partial t}\right) = a + i(P_g - C_H)V$$

a = optimal "soil rent" under long run plan

C_H = total harvesting and delivery cost

If today's price, $Pf_{(to)}$ is equal to the expected price in the next period, $E(Pg_{(t1)})$, the condition for maturity for a mature stand becomes:

(28)

$$E(P_{g_{t_1}} - C_H)E(V_{t_1}) - (P_{g_{t_0}} - C_H)V_{t_0} = \bar{a} + i(P_{g_{t_0}} - CH)V_{t_0}$$

where: $E(V_{t_1}) - V_{t_0}$ = expected change in volume between (t_0) and t_1)

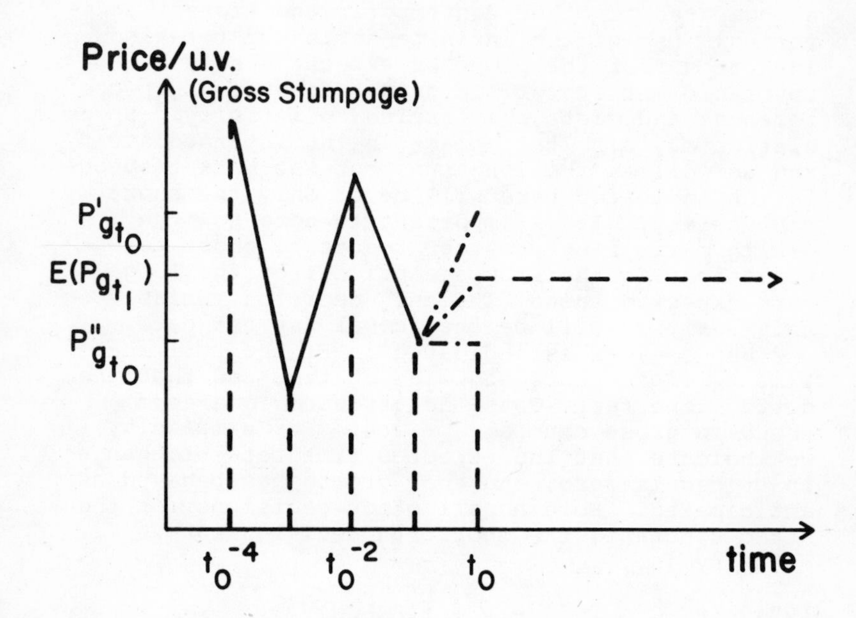

Figure 2.8

If today's price increases to P_{t_0}' as is indicated in Figure 2.8, the above stand is over-mature, holding other variables and parameters constant. Likewise, if the price today (see Appendix A.10) is P_{t_0}'', the same stand is under-mature. Hence, if the price in any current period is higher than the expected trend in prices, the producer has the incentive to harvest any timber that he would have harvested via his original management plan along with timber that is almost mature via the long run trend in prices. If the producer owns timber with a variety of ages, he will increase the rate of harvest when current price exceeds the long term trend in price and reduce the rate of harvest when the current price is less than the expected future price (next period). Given a timber inventory with some regularity to the distribution in age classes, when current price exceeds expected future price, the rate of harvest will exceed the rate of growth. On the other hand, when current price is less than expected future price, the rate of growth will exceed the rate of harvest.

C_H, the per unit of volume cost of harvesting and hauling is likely a functin of more than simply the volume harvested. As the amount harvested in a unit of time increases, the per unit harvesting costs will likely increase as well (see Alchain,]959). Overtime pay, constructing roads at a faster rate, wear and tear on equipment all tend to make C_H a positive function of the rate of harvest in any

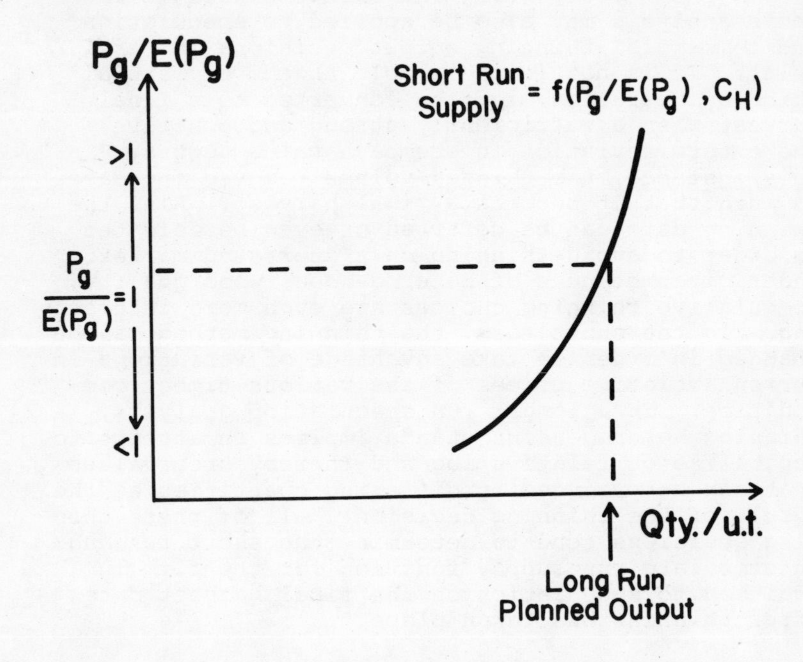

Figure 2.9

realistic analysis (not to mention the administrative costs of revising harvesting decisions). The effect of the change in C_H associated with different harvesting rates tends to make the short run supply for the firm increase at a decreasing rate. As is indicated in Figure 2.9, at $P_g/E(P) = 1$, the actual harvest is equal to the planned harvest. As P_g exceeds E(P) the rate of harvesting increases as does C_H, the per unit cost of harvesting. The short run supply under conditions of speculation is upward "sticky." The assumption of wealth maximizing behavior simply suggests that the firm will speculate on prices and harvesting costs in order to determine the appropriate output level at different points in time.

While the analysis of output under conditions of market speculation excluded thinning as a decision variable, the ommision was more for purposes of clarity than necessity. The market forces in the above analysis may also be applied to speculation and commercial thinning output decisions in particular. It is not inconcievable that a commercial thinning operation might be converted to a final harvest when a sufficiently strong and positive short term deviation in stumpage value occurs. In this instance, the thinned volume removed increases so much that it becomes a final harvest. Also the thinning date can be deferred or even be deleated in order to avoid thinning on a depressed market. Under circumstances of heterogeneous wood quality, speculative thinning choices are even more interesting. In the short term, the thinning method can be changed in order to take advantage of variations in current relative prices of the various timber commodities. In a longer run speculation analysis, thinning heterogeneous stands implies an attempt to capitalize on relative log and thereby tree values that can be produced in differing quantities as the result of the thinning decision. All of these thinning decisions tend to determine the short run and intermediate run supply function for the firm in addition to speculation on the final harvest date after thinning has taken place.

MEASUREMENT AND MEANING OF CAPITAL

In Chapter 1, there was a brief statement that both land and trees are capital. Yet the term capital often is used in confusing and differing ways. Hirschleifer (1970, p. 153) states that "at least three distinguishable meanings are common in the literature; they may be denoted as (1) real capital, (2) capital value, and (3) liquid capital." In fact all three meanings of the term capital apply to the timber firm.

Real capital is physical objects. Hence the quantity of land and timber measured in physical units such as acres and board feet constitutes the real capital stock of the forest. That the real capital stock can change over time can be easily observed in terms of land allocation, along with inventory, rates of forest growth, and removals.

What remains to be distinguished is the difference between liquid capital and capital value. Both are present in the Faustmann solution. If a forest landowner possesses bare land, the only value present is capital value. This is the present value of the infinite series of harvests. Where merchantable timber is established on the land, both the capital value of timber and the liquid value of timber are present. Liquid capital is the amount of current financial capacity available for investment. For any given stand of timber, it is the current price times the current standing volume in the inventory (PV). That wealth maximizing producers don't cut all of their growing timber as soon as there is positive liquid value attests to the fact that the present or capital value of the inventory exceeds the liquid value of the inventory and capital value of the land until the trees are financially mature. The Faustmann condition for harvest is one where the time change in the liquid value of the inventory or stand minus the intertemporal cost of holding the liquid value of the inventory is equal to the capital value of the land. In turn the liquid value of the forest is based upon the real capital and its current per unit price.

The definitions of capital are basic to determining what kinds of changes in capital occur over time. In a supply perspective, the definition of capital that is most important is that of real (physical) capital. In Equation 11 of this chapter, it was shown that a once and for all increase in demand increased long term supply due to higher levels of stand establishment. The average real timber capital per acre per unit of time increased. Increasing the average growing stock will unambiguously increase the total growing stock.

Efficient capital management implies changing the physical stock of capital, the physical rate of appreciation and the optimum rate of turnover of capital. The physical rate of turnover of capital has two measures. In terms of inventory, the rate of turnover is simply mean annual increment. In terms of land (physical), it is the number of years that an acre of land is devoted to a particular crop.

When one measures the rate of return on forest capital, all three definitions of capital are either explicitly or implicitly involved. Equation (5b) can be expressed in terms of (i), the rate of interest.

$$(5b)\quad \frac{P(\partial V/\partial t)}{a+PV} = i$$

where
V = optimum volume/AC at maturity
a = capitalized value of the site

Keeping in mind that the above is the marginal rate of return on forest capital, the condition is one where the ratio of the rate of appreciation of liquid capital over the sum of the standing liquid value of timber capital plus the capital value of land is equal to the rate of interest or economy wide marginal product of capital.

The definitions of capital are basic to other decisions. For example, it was just shown how short term price expectations would change the rate of harvesting of current standing inventory. In this instance, the capital value of land was constant and short term price fluctuations altered expected returns on liquid capital. In the next chapter, there will be discussion of long term adjustments in growing stock and the rate of harvest. It is essentially the changes in all three measures of capital that produces the dynamics of the firm and industry.

SUMMARY AND CONCLUSIONS

The nature of time often is troublesome in economic analysis. While time is not a factor of production in the sense of a physical input, observable timber production requires time. It is impossible to buy time as well. Interest rates are not the cost of time; but rather, exhibit a preference for the earlier as opposed to later availability of harvestable products.

Considerable attention was given to the timing of timber production decisions. The planner rightfully orchestrates a series of production and output decisions over time. Wealth can only be maximized when the right decisions are made at the right time. Likewise, the question of timing may at first glance appear as an independent factor of production. Timing more rightfully belongs in the domain of returns to the holder of contracts to all productive factors, namely the firm. Just as the rent was found to include more than the value of the fixed factor, the site, (Equation 5b), returns attributed to optimal timing must belong to the firm and subsequently add or subtract from its ultimate wealth position.

The manner by which time affects output was initially categorized as being of two kinds: One in which the time period of growth complemented the productivity of the variable factor; and the other where time was found to be a substitute for the variable factor. As the analysis expanded to n-decision alternatives, it was indicated that the usefulness of the distinction between time-complements and time-substitutes fell for all but marginal changes in the product price level.

Production, with the complete set of alternative inputs, input dates, and outputs dates variable, was indicated as the long run analysis of the firm. As the number of decisions held constant increased, the length of the run fell. However, all decisions are made in the light of their long run opportunity cost. Costs change instantaneously, but the long run optimum management plan may often require first vacating the site in order to implement it. With changes in parameters, the firm is continuously in equilibrium, given the opportunity cost of the then optimum long run plan.

Production and market speculation were examined in the intermediate and short run but all decisions were made again in light of the long run optimum plan. The definition of opportunity cost used is how any decision or event affects the wealth position of the owner of the production assets. This kind of analysis, incidently, varies significantly enough from the more traditional formulation of the firm in the long and short run with a static production function to merit its explicit attention.

It is anticipated that some readers will object to the exclusion of regulation of the growing timber inventory as a relevant economic consideration for the private firm. Regulation, or structuring the maturing stands so that they don't all approach maturity at the same proximity in time, has been perhaps the overriding concern in American Forest Management (see for example, Davis, 1966). Normally, a management regime is imposed on the entire forest holding and then harvesting proceeds, holding the management regime constant in order to eventually provide an age structure (taking account of differences in the productivity of various sites) that will yield a continuous flow at the predetermined harvest dates. The material on speculation suggests some of the economic problems associated with continuous flow and variations in production and prices. There appears to be some fundamental problems associated with the traditional forest regulation concepts. If the firm desires to structure its inventory in some way, it can accomplish this end by varying the input decisions as well as the harvest date. Furthermore, would the private entrepreneur desire to regulate his holdings if the composition of his inventory complemented the inventories of his competitive counterparts? It seems reasonable to believe that the owner might actually desire to regulate his stands if only to ward off the risks of poor market conditions at maturity, but the literature appears to have largely overlooked this consideration. If quantity supplied is low due to some vagaries in the age-volume structure of the total inventory in existence, current price would exceed the expected future price and harvesting slightly younger stands would commence. The same market determination of output occurs with any bulge in the quantity of mature timber.

It is also believed that a well functioning market may be able to accomplish a "regulated" forest. While the proof is beyond the level of this work, the basic elements of supplyand demand over time coupled with price differentials associated with intertemporal supply-demand imbalances provides the essentials for such an effort.

NOTES

1. For a clear exposition of this point with a

linear homogeneous production function, see Ferguson (1969, 133-137).

2. If the slope was less than one, planting outlays when compounded would overwhelm stumpage returns.

3. Unlike static production with two factors, the cross partial derivative can be positive. The solution for the firm is still stable because the opportunity cost of the stand increases exponentially over time.

4. A similar approach to this can be found in Goforth and Mills (1975). However, they don't appear to recognize that a fixed investment duration will not minimize wealth, nor do they explicitly constrain their solution by an infinite series of equal rotations.

3 Theory of the Timber Market

INTRODUCTION

In order to appreciate the unique attributes of market determination of timber investments, output, and price, it is helpful to first briefly review the neoclassical theory of the competitive firm and

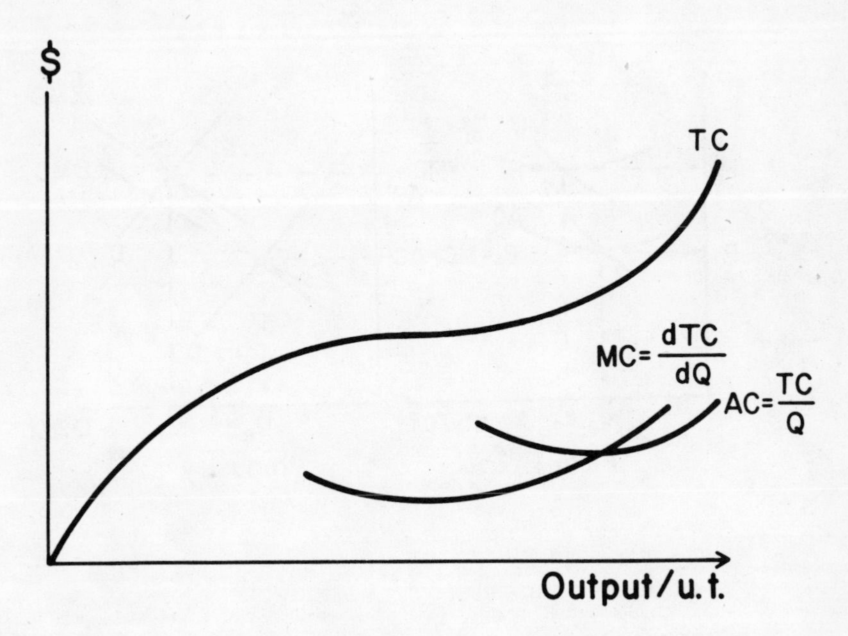

Figure 3.1

its relation to aggregate market supply. The firm's long run total cost function is defined as the locus of the sum of all expenditures per unit of time for a variable quantity produced per unit of time. It is implicit in this analysis that each point on the total cost function represents the most efficient allocation of factors of production where factor costs are constant and all factor quantities are variable (Henderson and Quandt, 1958, pp. 55-62). As Figure 3.1 indicates, total cost increases at an increasing rate throughout the efficient range of production (i.e., diminishing returns to the factors of production). Long run average cost per unit of time is simply total cost divided by quantity and is characterized by the typical U-shape. Marginal cost or dTC/dQ intersects average cost at minimum average cost.

The long run equilibrium point of operation for the firm is represented as point q_e in Figure 3.2a. At this point, marginal cost (MC) equals average

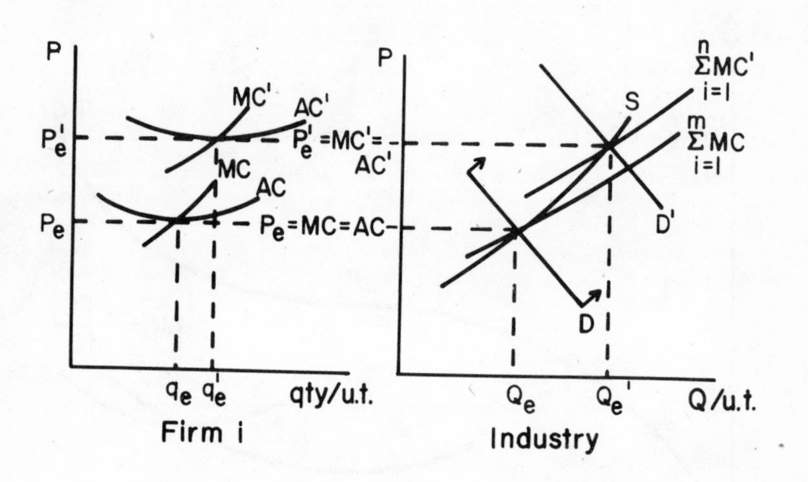

Figure 3.2a Figure 3.2b

cost (AC) equals marginal revenue (MR) equals average revenue (AR). At point q_e, either the profit maximizing or cost minimizing competitive firm has no incentive to change output and is operating at a point where all costs are covered.

The aggregation of firms into a competitive industry is represented in Figure 3.2b. with the common postulate of a negatively sloped demand function, the sum of the marginal cost functions for all operating firms intersects market demand at equilibrium price P_e. At price P_e and quantity q_e, total industry output is the sum of the outputs individual firms are willing to offer at the equilibrium price.

If industry demand should shift to D' in Figure 3.2b, each firm attempts to expand its output along its respective marginal cost function or expansion path. In the process, demand for the scarce factors of production increases. With any conditions short of perfectly elastic supply for the productive (normal) factors, expanded industry output will increase factor prices. Higher factor costs coupled with the potential for entering new firms shifts the sum of the marginal costs function in Figure 3.2b to $\Sigma MC'$. The industry supply function (SS in Figure 3.2b) is the locus of equilibrium outputs for various levels of price allowing the number of firms and factor prices to vary (Friedman, 1962). The equilibrium supply function is not the sum of the marginal cost functions under the above conditions. Rather, the equilibrium supply function intersects the industry marginal cost function from below at the equilibrium price and output.

This brief introductory review of the static theory of supply lays the grounds for a considerable amount of material covered in this chapter. It serves to start the discussion on more familiar grounds before undertaking the new terrain. Just as market demand affects factor prices in the static analysis of the market, the same will be true, for example, in the case of time dependent timber production. The market will solve the optimum time length of growth along with other investment decisions, thereby determining market output and price.

THEORY OF PRIVATE TIMBER SUPPLY WITH TWO DECISION VARIABLES

Of the two variable models presented in Chapter II, the one in which time and establishment are substitutes is more restrictive in terms of establishing a positive slope to the firm's supply function. Here, that model is used as the basic approach and is later contrasted with the two variables complements case.

It was proven that if the initial mean annual increment was greater than a trivial level of one, the supply function of the firm was positively sloped (Equation 11). The initial assumptions, as well as Appendix A.1, indicated that the marginal firm is indifferent between an infinite series of timber harvests and some other valued productive land use. The marginal firm is not necessarily an occupant of the least productive land. Rather, the marginal firm's rent is closest to zero in the sense of value in excess of opportunity cost in non timber use. (Lerner, 1944). Changes in the supply of land were indicated to be simply the entry and exit of new firms. As the value of timber rises or falls along with the direct costs of producing or harvesting it, wealth maximizing behavior suggests that the economic definition of marginal land changes and new firms will enter or existing firms will exit the industry.

With this information, it is possible to sketch a first approximation of private timber supply. In Figure 3.3a, the representative firm's contribution is denoted by the optimum mean annual increment per acre at various prices, times the number of acres owned by the firm. The firm's supply function represents the optimum management plan for various price levels holding the interest rate (i) and per unit cost of establishment (w) constant. In order to maintain a degree of analytical simplicity, the assumption is made that the timber purchaser bears the hauling and harvesting costs as a trade practice. Therefore, the price axes in both Figures 3.3a and 3.3b are net stumpage prices for the homogeneous product. The firm's contribution to market supply is strictly analogous to the firm's static marginal cost function alluded to above. They are both expansion paths at constant levels of the non product price market parameters, (i and w), with all production decisions variable.

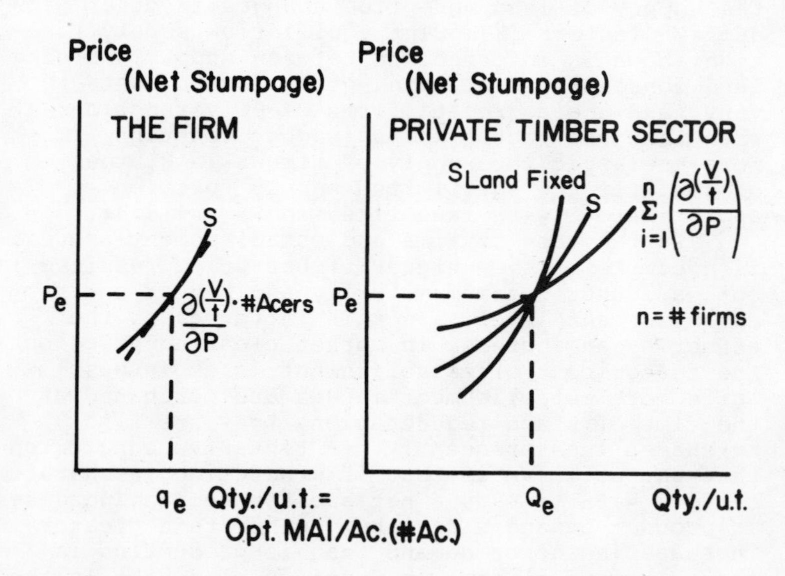

Figure 3.3a Figure 3.3b

In the aggregation of private sector timber supply, the sum of the individual firms' supply functions is strictly analogous to the sum of the marginal costs depicted earlier in Figures 3.2a-b. Both are aggregated under conditions of constant market parameters other than the product price. In Chapter II, it was indicated that as the net price of stumpage (P) increased, more established inputs (X) were purchased more frequently for the substitutes case. As all private producers attempt to purchase more stocking factors, the factor price (w) will increase as long as the price elasticity of supply is not infinite. Entry and exit of new firms also affects the slope of the aggregate timber supply function as well. In order for new firms to enter the industry, the value of marginal and sub-marginal land must increase in the timber production alternative. (This process has apparently taken place in the Southeastern United States where much of the private timber land was previously allocated to cotton production and other agricultural uses.)

Thus, the equilibrium supply function (S) in Figure 3.3b, must be influenced by both changes in the supply of land and price changes in other productive factors (X). The equilibrium supply function shown is intermediate between supply, holding land constant but allowing other factor costs to vary, and the aggregate firms' supply functions with land and factor costs (wages) constant. The more inelastic the supply of timber land, the closer will the equilibrium supply function approximate S with land fixed in Figure 3.3b.

In the case of time and establishment as decision complements, a higher timber price resulted in both a longer optimum maturity and higher optimum establishment level. In this latter case, the effect of an increase in market clearing price on the factor cost of establishment is ambiguous. While more establishment inputs are purchased at the time of stand regeneration, they are also purchased less frequently. A tentative conclusion that can be drawn is that if production is characterized with time as a net substitute, an increase in product demand will likely generate a greater increase in factor demand than if production is characterized by time as a net complement. In the former case, factors are bought more frequently and in the latter case, the frequency of purchase by the individual purchaser diminishes. Given diminishing returns to the establishment input in the variable complements case, it appears to be intuitively correct to suggest that at higher establishment levels, an increase in price will induce an increase in the factor cost of establishment. In this latter instance, the establishment input would again be a "normal" factor of production. Higher product price would induce an increase in the demand for productive factors.

In the more general n-decision variable case, where the number and frequency of inputs such as fertilizer and the quantity per unit of time of each input are all variable, it appears as though the distinction of time as a net substitute versus a net complement is not as important. Even if a higher price results in a longer harvest age, in general, the expanded management intensity with n-decision variables can increase the number of interventions and/or the quantity of the input utilized per intervention. A cautious conclusion that factor prices will increase as the private sector attempts to expand, regardless of whether time is a net substitute or complement, appears warranted. In this sense, then, we are excluding a form of inferior factors of production.

The establishment factor would have to be classified as an inferior factor of production under the following circumstances. An increase in timber demand generates a longer optimum growth period and results in a reduced frequency of establishment factor purchases. If the reduced frequency of purchases actually offsets the higher quantity purchased upon each harvest, the combined changes in quantity and frequency of purchase result in a reduced factor demand.

SUPPLY IN THE PUBLIC SECTOR

A brief review of some of the operational aspects of sustained yield forestry as generally practiced on federal lands in the United States is presented. More intensive appraisals such as Waggener's (1969), exist for further reference to interested readers. Of greater importance to this research are the bureaucratic interpretations of the sustained yield philosophy and legislation which affect the equilibrium quantities forthcoming from the federal (public) lands at various prices.

One critical aspect of private timber management is the determination of the harvest age. Typically, in even-aged management, a yield table for a "fully-stocked" stand is used to determine the age of maximum mean annual increment. Given homogeneous products, no expected change in real prices, and constant harvesting costs per unit of volume, an infinite series rotation with a zero discount rate would give a rotation age at maximum mean annual increment. In almost every instance, the financial determination of the optimum time to harvest is considerably shorter than the maximum sustained yield policy. (See Gaffney, 1957).

The existence of these extended harvesting ages under sustained yield management in the public sector reduces the justifiable management intensity. At an annual interest rate of 6%, the marginal dollar per acre invested in year one must pay at least $10.29 in year forty; and at least $105.80 per acre in year 80 in order to be warranted. For the most part, stand values with homogeneous products do not increase at increasing rates between the ages of forty and eighty. Thus many investments that can be warranted with financial determination of the harvest are frequently unjustifiable when properly analyzed under current public sector policy constraints.

The long public rotation age policy has certain implications regarding output responsiveness to different price levels, both in the short and long term. It takes a much greater change in per unit stand value to justify an increase of an additional dollar invested in year one with an eighty year rotation than it does with a forty year maturity. As a result, the long run supply function of the public sector is less positively sloped (less price responsive) than that of the private sector. This should not imply that all public sector production decisions are made consistent with sound investment criteria. In fact, the current analysis by the U.S. Forest Service of the "Allowable Cut Effect" represents a means of inflating the desirability of budgetary requests for investment projects that are highly suspicious under careful scrutiny. (Sassaman, Schallau, Schweitzer, 1973; Teeguarden, 1973).

Legislation governing the management of the national forests requires that timber harvests be conducted in a nondeclining, even flow manner. Just how even the flow is has been questioned by Johnson (1977). From a supply perspective, the more invariant (even) the short term supply, the less elastic will national forest supply be with respect to price, production costs, interests rates and so forth. A discussion of some of the research implications of national forest supply policies is found in the final chapter.

THEORY OF TIMBER DEMAND

Demand for timber is generated from two principal and often opposed kinds of use. Timber often is a factor affecting psychic income through the enjoyment of an outdoor setting. In this instance, timber itself is a component of a final product; the outdoor or visual experience. The other use of timber is as a factor of production of wood products. Harvested timber, or standing trees which produce products such as resins for turpentine, all share the common characteristic that their demand is a derived demand based on the demand for and production of final goods. The second type of timber demand is the exclusive subject of this section. By concentrating attention on the derived demand for timber, it is possible to isolate the theory of timber production from the multiple-use issue.

It is possible to further categorize the nature of derived timber demand. On one hand, stumpage may be converted into durable capital goods such as houses or containers, and on the other hand, timber may be employed in manufacturing non-durable goods such as paper products. This distinction will come into use in the subsequent market analysis.

If timber were converted into a single product, estimating the derived demand for stumpage would be a relatively easy econometric project. However, current forest products industry technology encompasses the production of multiple products even at the same mill. A firm may process one log into veneer, lumber and chips, and then even burn the remaining residue to produce power. Often, no explicit value breakdown of the various portions of the log in alternative uses is made in the market place. Furthermore, it is technologically possible to vary the proportions of the same log in the allocation to various products.

The total value of the log and thereby the timber stand, is the sum of the highest subcomponent alternative values. Each use is a different factor of production in manufacturing different products; chips in particle board or paper, veneer in plywood, and the sawlog portion for lumber. Actual log utilization is an intra-firm allocation decision. While firms may exchange logs, the actual milling decisions are made by the firm and are hence non market in nature. Market information regarding the factor values of the various portions of the logs is sketchy, to say the least, and production itself is characterized by joint products.

In order to estimate derived stumpage demand, production functions for the various products would have to be known. Alternatively, the costs of all productive factors could be used to estimate the production functions. However, even with the best data, estimating derived stumpage demand is a complex empirical problem and well beyond the scope of this study. Yet, development of an empirically based forest products production function is necessary in order to predict an unbiased estimate of timber demand.

In the absence of a reasonable method of estimating an unbiased timber demand function, economists often turn to the factors which influence the final demand for products. The view, and correctly so, is that timber demand is an implicit function of the demand for final products. Economists dating back to Marshall have recognized the relationships between the derived demand for a factor of production, availability of substitutes, demand for final products, and the elasticity of supply of substitutes. (See, for example, the discussion in Friedman (1962) under fixed coefficients of production.) If, however, one excludes supply characteristics of other productive factors, it should be recognized that demand estimates are biased by misspecification or under specification. (See Rao and Miller, pp. 32-34).

This introduction should both point out some potential hazards in estimating timber demand and serve to delineate the limitations of the discussion on timber demand. The approach used here is to list only variables which affect the demand for final products of which wood fiber is a major component. The variables listed are believed to be the most important of a much larger set. Only the demand variables which are viewed to be of long term importance are listed in equation (28). Short term demand considerations are treated in a subsequent section.

(29) $Q^D{}_{stum/capita} = f(P_{stum}, Y/capita, i, Tastes, Age)$

where $Q^D{}_{stum/capita}$ = per capita quantity of stumpage demanded.

P_{stum} = real stumpage price

Y/capita = per capital real income

i = real rate of interest

Tastes = tastes and preferences of the population

Age = age structure of the population(% of pop. age 20 to 30)

It should be noted that no variable indicative of the prices of related goods is specified. Working with the variables which affect final demand for wood fibre products precludes the specification of such factors as labor costs and milling capacity.

Holding tastes and preferences constant, economic theory suggests an equation with the following signs for the independent variables.

(30) $Q^D{}_{stum/capita} = \alpha_1 P_{stum} + \alpha_2 Y - \alpha_3 i + \alpha_4 Age$

Ceteris paribus, quantity demanded is a negative function of price. The present value of durable goods is inversely related to the rate of interest. Hence, quantity demanded for raw material in the manufacture of durable goods is an inverse function of the real rate of interest (Witte, 1963). Holding other things constant, the per capita quantity demanded is also a positive function of per capita real income. Finally, the age structure of the population in terms of the initiation of new households positively influences the demand for timber. This limited list of variables has been specified for inclusion in the next section on the comparative statics of the private sector.

MARKET EQUILIBRIUM IN THE PRIVATE SECTOR

Throughout the remainder of this chapter, the attention rests exclusively on determination of price and output in the private sector. The subject of public timber output and its role in the market place will be reintroduced in the chapter on forest taxes.

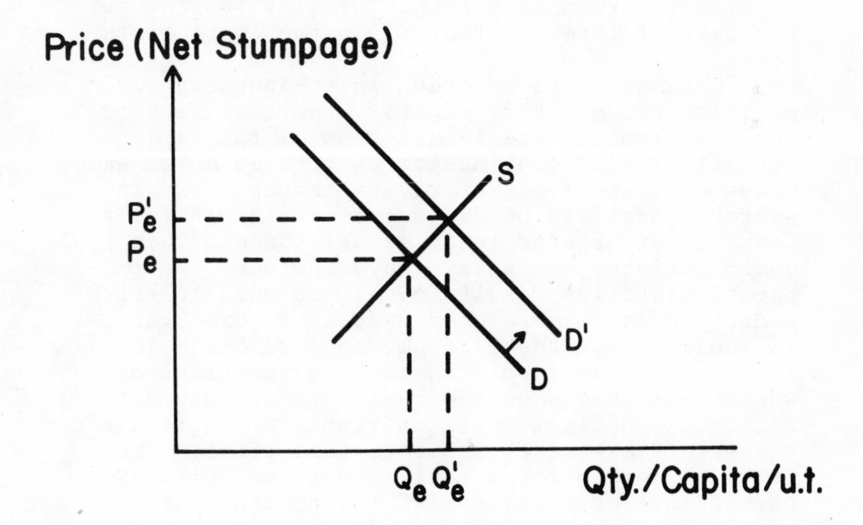

Figure 3.4

Figure 3.4 represents initial market equilibrium at output Q_e and price P_e. Price will again be net stumpage price with costs of harvesting and hauling being the liability of the purchaser. We might first investigate the effect of a shift in demand arising from a shift in tastes and preferences, per capita income, or the age structure of the population. When D shifts to D' in Figure 3.4 in a once and for all manner, the market clearing price which each firm takes as a parameter increases. Where the production function in the two variable case is characterized by decision substitutes, firms shorten their rotation ages and increase the optimum establishment (Equation 11 Chapter 2). In addition, to the extent that the supply of timber land is price elastic, new firms (land) enter production. Output increases until the market is again equilibrated at the higher price P_e' and expanded output Q_e'. Had the production function been characterized by decision complementarity, establishment and optimum maturity would have both increased (section G, Chapter II). As was indicated earlier in this chapter, the expansion of output increases the factor cost of establishment as more of the physical inputs are demanded per unit of time (ruling out the case of inferior factors in the complements case).

Because there is great inter-industry competition for monetary capital, the real rate of interest (money rate less a premium for expected inflation) will continue to be treated as an exogeneous parameter to the private timber producing sector. Analysis of exogeneous shifts in (i) are possibly of greater intellectual concern than real world activity warrants. Nevertheless, the comparative statics results of a once and for all reduction in (i) are presented in Figure 3.5. As is indicated in the diagram, the reduction in (i) shifts both the supply and demand functions outward. Not only does the real rate of interest affect the optimal level of timber production and thereby timber supply, but it also affects the level of demand for durable goods (equation 29). As the interest rate falls, the present value of the stream of productive services yielded by the durable goods increases, and thereby the demand for durable goods increases. Where the demand for timber is derived from the demand for houses and other capital items, one would predict that a lower

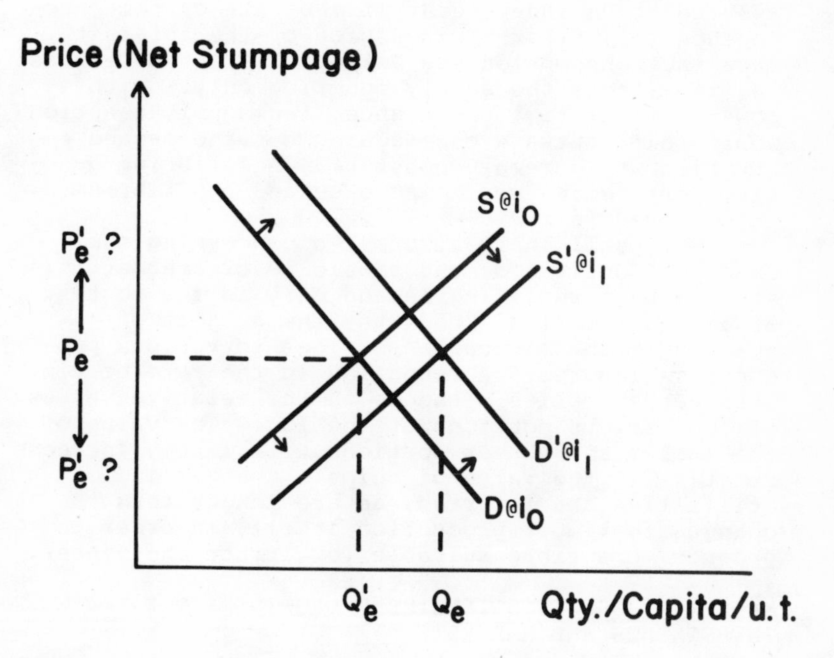

Figure 3.5

rate of time preference would result in a shift (increase) in the timber demand function (Witte, 1963).

The other unique circumstance in the stumpage market with demand derived from the demand for durable goods, is the unpredictability of the direction of movement of the equilibrium stumpage price. Whether or not price increases, depends on the respective magnitudes of the shifts in both the supply and demand. Both quantity supplied and quantity demanded are functions of the rate of interest (i). As is indicated in Figure 3.5, a reduction in the rate of time preference shifts the supply function (section D, Chapter II and Appendix A.5) in the same direction as the shift in the demand function. The equilibrium quantity unambiguously increases with a reduction in (i) but the direction of the movement in price is not known.

Suppose now, that the demand for timber had been derived excusively from the demand for non-durable products such as disposable paper products. Under these circumstances, quantity demanded is assumed to be independent of the rate of time preference (i). (For a discussion of the interest rate and consumption see Bailey, 1971). A decrease in (i) affects the supply function only. With a lower rate of time preference, the supply function would shift outward once again, but the demand function would remain constant. Equilibrium quantity forthcoming would increase and equilibrium price would fall.

The preliminary discussion concerning the theory of derived demand should be clearer at this stage. By formulating demand only in the context of demand for final products, the subject of changes in the allocation of logs to various products is ignored. A reduction in the rate of time preference implies a change in the relative values of the various portions of the log. The value of the lumber and veneer portions will likely increase relative to the value of chips. Changes in log utilization are inferred, as are longer term changes in timber production itself, in order to produce more fibre suitable for lumber and veneer.

MARKET ADJUSTMENTS IN TIMBER GROWING STOCK, INVESTMENTS AND HARVESTS

Thus far the analysis has concentrated largely on the analysis of steady states. In order to develop the concepts of flow adjustments, or the actual rate of harvest, we must first look at the relationship between harvest and growing stock. Recalling that timber is mature when the time change in value equals the storage costs of inventory (including land), the long run steady state model is one where harvest equals growth. With the simple point input, point output model, the harvest liquidates all of the accumulated growing stock on the site.

Where the forest inventory for all firms includes stands of different ages, the actual yearly harvest will be less than the actual cut. The relationship between actual cut, growing stock and uncut inventory can be expressed in the following identity.

(31) $AC \equiv GS - Q_R$ Where: AC = actual rate of harvest

GS = Total growing stock of all private owners

Q_R = Reservation demand (stock)

Of the three variables listed in the above equation, Q_R, the reservation demand is perhaps the least familiar to foresters. Simply stated, reservation demand is the quantity of the forest inventory which owners desire to reserve for growth and future harvests. While the reservation demand is actually a function of several variables, for purposes here we shall simply indicate that it is a function of the current net rate of appreciation of each stand included in the aggregate growing stock. This rate of value increase depends upon previous levels of establishment for each existing stand, the age of each stand as well as the levels of i, P, and w.

For any existing stand that is mature, the relationship can be expressed by manipulating equation (5b) from chapter two as follows:

$$(5b') \quad P\{\partial V/\partial_t (\overline{X},t) - iV(\overline{X},t)\} = i\{\frac{PV(X,t)-iwXe^{it}}{e^{it}-1}\}$$

where $\overline{X}$ = level of stocking of existing stands fixed by previous decisions

We can readily see that if the price of the existing inventory shifts to some higher amount, the mature stand indicated in equation (5b') is over mature. Multiplying the price by a constant K, where K is greater than one, increases the value of the growth medium, or land, relative to the net rate of appreciation of timber even before the new optimum plan is designed. This is shown in equation (5b'').

$$(5b'') \quad KP\ (\partial V/\partial_t - iV) < i\{\frac{KPV-wXe^{it}}{e^{it}-1}\}$$

As a result, timber in the current inventory will have a shorter rotation age than was originally planned. Firms will want to liquidate their timber at a faster rate in order to replace it with optimum stands. The result of a shift in demand (increase) and resulting higher prices on both reservation demand and actual cut is shown in equation 32.

(32) $\partial GS/\partial P > 0, \ \partial Q_R/\partial P < 0.$

As was shown previously, higher prices also increase the aggregate level of investment. The long term adjustment in growing stock also increases. The adjustment in initial growing stock, actual cut, and long run supply are indicated in Figure 3.6.

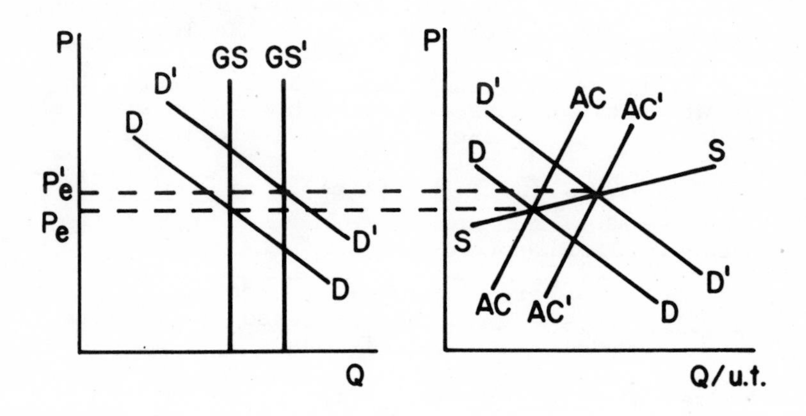

Figure 3.6

In the lefthand side of the graph, initial growing stock (GS) is intersected by the market demand curve (DD) at equilibrium price P_e. In the right hand side of the graph initial actual cut (AC) intersects demand at the same price. If market demand shifts to D'D', initial cut will expand along the short run supply function (AC). However expansion of investment in timber will gradually expand the stock of timber to (GS') thereby shifting the short run supply function to (AC'). The long run supply function is SS and is synonomous with the theory of long run supply previously elaborated upon.

Recalling the two theoretical relationships between (X) and (t) (substitutes versus complements), there are also two different adjustment scenarios. When establishment and time are substitutes, the adjustment is fairly straight forward. Cutting the initial timber growing stock at a faster rate and re-establishing timber with both shorter harvest ages and higher annual growth rates allows for a continuous supply over time.

However, where the production function is characterized by two variable complementarity, a once and for all increase in the demand for stumpage presents a different kind of adjustment problem. On one hand, the incentive exists to replace current inventories at a faster rate. On the other hand, the newly established stands have a longer optimal duration. The question arises as to whether actual cut (AC) will be time continuous during the adjustment period.

The point can be demonstrated with an example. Suppose that the initial optimum duration is 45 years. Demand shifts so that subsequent stands are to be grown 50 years. If owners want to remove the initial inventory at a faster rate and then replace it with growing stock with longer harvest ages, some form of adjustment will be required that allows the timber to be withheld from the market during the adjustment between steady states.

A brief tangent to the static theory of the firm provides an interesting comparison. With a static two factor production function, the factors must be substitutes in order for the firm to achieve a finite profit level. (See Quirk, 1976, p. 115) In the case of the timber firm, a finite wealth maximizing position is possible when time and establishment are complements. Whether or not market supply will both be positively sloped and continuous over time is an unresolved matter. The possibility exists that the incentives to age inventories further by withholding supplies could increase price furthering the incentive to age inventories and so forth leading to an unstable type of reaction function under conditions of competition. (See Henderson and Quandt, 1958, pp. 110-124)

Certainly the adjustment issues are exacerbated with the assumption of two variable complementarity. In the more complex production functions, supply responses are likely to be both faster and stable. Additional inputs can be added to already initiated stands and thinnings can be rearranged, during the adjustment period. Even though time of maturity may be a net complement, it is believed that the question of stability is of more theoretical interest than real world concern when viewed in the context of realistic production alternatives.

MARKET SPECULATION

Considerable attention was given in the previous chapter to the theory of speculative behavior on the part of the firm. Input and output decisions were based on the divergence between current and expected future prices. Short term speculation was seen as the behavior of the firm under conditions where there was variation between current price and expected future price but where the expected rent attributed to the employment of the fixed factor was constant.

One of the most volatile variables affecting timber demand is the variation in relative mortgage rates as loanable funds are bid away from or move toward potential home purchasers (Alberts, 1962; Downs, 1974). The overall speculative behavior of firms aggregated into the private timber sector is indicated in Figure 3.6. Again the demand for stumpage is derived from the demand for durable goods (homes). As the mortgage rate falls relative to another borrowing rate such as the rate on some type of long term corporate bond, people desire to purchase homes and the demand for timber increases. On the other hand, as corporations expand their level of investment, funds are bid away from the domestic housing market and the mortgage rate increases relative to the corporate bond rate. This formulation of housing investment indicates that housing activity runs countercyclically to the rate of business capital accumulation.

In figure 3.7, total planned market output is indicated at point Q_e, the point at which current price is equal to expected future price. As the mortgage rate falls relative to the corporate borrowing rate in the current period, the demand for housing increases, more housing starts are ini-

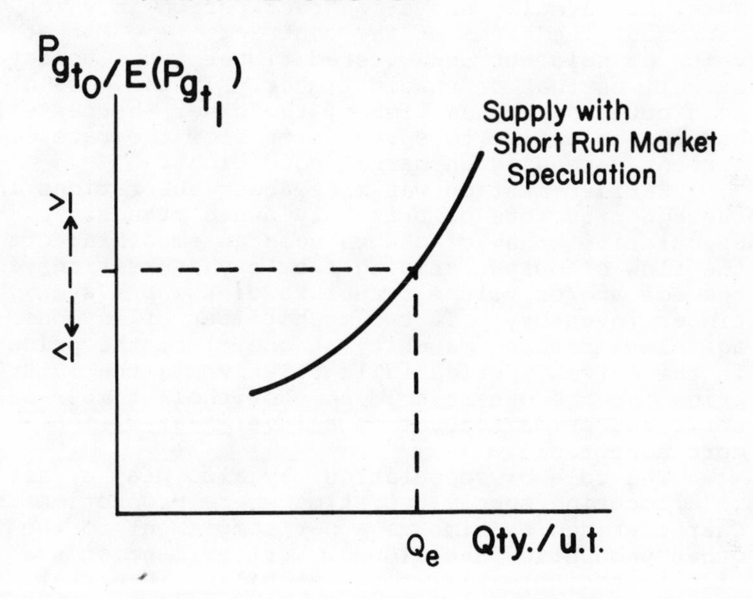

Figure 3.8

tiated and the current price for timber increases relative to the expected future price. All of the planned harvest (Q_e) is taken during the period, along with some additional timber that would have been slightly less than mature at a lower current price. If, on the other hand, the mortgage market turns in a manner which reduces the incentive to initiate new homes, the rate of output falls below expected future price and timber is stored "on the stump" until demand returns to a level which warrants a greater rate of harvesting.

Notice that the output expansion path as a function of current price relative to expected future price is indicated as being upwardly sticky, (Chapter 2, Figure 2.9). As the rate of harvesting expands, more overtime and logging equipment is required which reduces the profitability of further expansion. The diagram indicates that the relative prices are once again net of harvesting costs. This does not preclude the buyer from being a speculator as well as the seller. If the buyer is responsible for the hauling and harvesting of a sale, he normally has a contract period in which to carry out the terms of the contract. The market value of sold but unharvested timber can fluctuate as well as that of unsold timber. In the case of sold but unharvested timber, the owner (harvester) has the incentive to speed up or slow the rate of harvest depending on market conditions.

Earlier mention was made about aberrations in the age structure of privately owned stands. Speculative behavior can be seen as smoothing out the flow of output arising from any aberrations in the age and/or volume structure of the private timber inventory. If too much timber of any one age class reaches maturity at one time, the price in the current period falls relative to the future price and the owners hold some of their timber back until the price returns to a level that will be more acceptable.

The role of speculation may also play a part in allocating supply over time where production is characterized by time as a net complement to the other production decisions. Earlier mention was made to the potential for instability in market supply over time. If owners attempted to withhold supply from the market in order to achieve a longer optimal maturity as part of a market adjustment, the current price would increase relative to the future price. Current output would again be forthcoming in a time continuous fashion.

SUMMARY

The major focus of this chapter is on the market solution of the level of private timber investment and output. While the development of a market with speculation has resulted in a theoretically stable solution, (i.e., the supply function has a positive slope) many practicing foresters still view private timber production with market allocation of productive factors with great distrust. The investigation of the market under private ownership has not yielded any theoretical basis to substantiate continued belief in a "timber famine" under private production. In fact, we find that land and other productive factors are allocated efficiently both within the industry and between industries in the private timber market.

Certainly, private timber production involves planning over periods that are uniquely long in American industry. Personal discussion with a representative of one of the major producers indicates the use of modeling techniques which predict the results of various combinations of managerial intervention on cost and output forty to eighty years hence.[1] Unlike the public sector counterpart, greed, the motivating force of capitalism, makes stumpage available as long as current and future market conditions warrant its production.

NOTES

1. Interview with the long range planning group at the Weyerhauser Company, (April, 1974).

4 The Problem of Forest Taxation

INTRODUCTION

This chapter develops propositions concerning various kinds of forest taxes and their respective influences on economic efficiency. Previous material on the timber firm (Chapter 2) and industry (Chapter 3) is utilized in order to indicate how different kinds of taxes:

1. affect the allocation of productive factors within the firm;
2. affect the firm's supply;
3. affect industry supply and price;
4. affect the allocation of land to timber production.

Additional, and rather brief remarks, are made concerning tax induced changes in the utilization of stumpage. Because the analysis in this chapter uses two decision variables within the firm, the scope of the study is somewhat broader than selected previous publications on the subject. The results of the tax analysis are reflective of state tax policies since taxes are assumed to leave the effective rate of interest unchanged.

GENERAL ISSUES

Economic research on taxation dates back to classical economists such as Ricardo (Musgrave, 1959). A distinction was made early in economic studies of taxation between the impact as opposed to the incidence of a tax (Seligman, 1921). The impact of a tax is the legal liability for and payment of the tax revenue (who remits the tax and how much is remitted). Tax incidence is a broader concept inclusive of the impact but in addition includes the shifting of the tax burden to others not liable for tax payments. For example, while one party may remit the tax, he may shift a portion of the remittance burden to others due to changes in product prices and quantities offered for sale. The tax incidence framework is used in this study of forest taxes.

Tax incidence is closely linked to the concept of economic efficiency. Taxes imposed on one industry which are shifted to individuals or other industries may result in inefficient allocaiton of resources both in the initial industry and in the other sectors to which the tax burden is shifted. No doubt, few taxes, if any, have ever been adopted using economic efficiency as the sole standard of justification. However, economic efficiency may be the only objective criterion for taxation analysis in the absence of an objective social welfare function (Arrow, 1951). The tentative nature of this statement should be made clear. The avoidance of value judgements, particularly in the area of distribution, is a value judgement in and of itself (Little, 1949). Nevertheless, because efficiency in production is one necessary condition for the attainment of maximal social welfare (Bator, 1957), any tax which reduces the level of economic efficiency by misallocating scarce productive factors, supercedes the attainment of the optimal production frontier.

The welfare evaluation of forest taxes implies another issue in addition to that of efficiency. Welfare implications concerning rent or the surplus payment to a productive factor beyond its opportunity cost in other uses, date back to Marshall (Mishan, 1964). What then of forest rent for the non marginal firm? Rent is either converted into current goods (consumption) or future goods via productive transformation (investment) given the competitive nature of the capital market. While rent can be altered by forest taxation giving rise to the changes in investment and consumption decisions resulting from changes in the net of tax economy wide rate of interest, changes of this nature lie beyond the partial equilibrium analysis context.

However, given an economy with a public sector which desires to produce an additional pure public good (Samuelson, 1955) with a positive present value at the competitive rate of time preference, forest rent may imply inefficiency. If forest rent can be taxed in such a manner so that timber supply and price is unaffected and the tax revenue is used to finance a pure public good with a positive present value, forgoing the public good and necessary tax represents inefficiency. Society could be better off by providing the public good without being worse off in terms of timber. If a tax only reduces the economic surplus or rent, without affecting product price and output of final goods, the tax represents a potential policy tool for efficiently providing public goods. Some distortion is likely inevitable. In realistic terms, the choice is often the provision of a given level of government revenue with the least output or efficiency costs arising from an alternative set of fiscal measures. The same choices are available where the goal of taxation is income or wealth redistribution.

Existing tenure arrangements regarding forest taxes and the timber industry are of utmost importance. Land and production in the public sector are not taxed in any manner[1] while private lands and forest production are subject to taxation. As such, private timber companies often incur operating expenses resulting from forest taxes which have no counterpart in the accounts of public timber firms such as the U.S. Forest Service, Bureau of Land Management, and various state and local agencies. Hence, forest taxes levied on the private sector of the industry, imply public timber supply responses due to tax induced stumpage price changes. If the steady state forest capital management regime in the public sector (Chapter 3) is suboptimal, a tax which shifts the competitive advantage toward the public sector will further magnify misallocation of productive factors as was alluded to in the last chapter.

Often in the economic analysis of taxes, tax jurisdiction is ignored. Yet, there are national forest tax policies and state and local tax policies. While there is a great variety of specific state forest tax laws, the tax instruments used are fairly limited. The analysis of national fiscal policy should likely entail the examination of additional variables than is the case of state policy. In the national setting,policy changes can easily alter the economy wide marginal product of capital. In the analysis of state tax policies, the shifts in investment are likely small enough so that changes in the real rate of interest are negligible. To the extent that state tax policies directly affect timber revenue, they can produce new supply-demand equilibra on a regional basis with secondary impacts on national outputs and price.

THE IMPACT OF FOREST TAXES ON THE FIRM

The discussion in the previous section should serve as a background for the analytical framework focusing on the impact of forest taxes. The theory of the firm with two decision variables (X, t) presented in the second chapter and the theory of the industry presented in Chapter III are the basic components of the analysis. The timber industry is composed of both a private and a public sector. Taxes are levied only on the private sector and it is assumed that taxes do not influence the public supply function. Land and factor costs are variable in the private sector so that it is possible to indicate how forest taxes may be shifted in terms of land use, timber price and the quantity of establishment inputs purchased.

Due to the previously mentioned view that the private timber sector is a special case for tax consideration, the analysis once again is partial equilibrium in nature. Initially, the timber market is in a steady state equilibrium. The equilibrium is then disturbed by a change in taxation policy in the private timber sector. Only marginal tax changes in the private timber sector are indicated for each kind of tax examined.

Several different kinds of taxes, all of which are close approximations of forest taxes employed in different states, are examined. The first tax to be examined is an excise tax. In forestry this tax is usually referred to as a "yield" tax. Essentially, the firm is liable for a percentage (R_1) of harvest value. Hence, the net of tax income from harvest is expressed as $(1 - R_1)$ (PV). A second kind of tax is similar to our capital gains tax. Under this tax, planting costs are capitalized and deductible from point output receipts. Where (R_2) is the rate of capital gains tax, net of tax income is; $((1-R_2)(PV) + R_2 wX)$. In contrast, an ordinary income tax defines taxable income as the difference between receipts and expenditures in any tax period. Defining (R_3) as the rate of income tax, taxable income is $(1-R_3)(PV-wX)$.

All of the above taxes are dated on the basis of the timing of receipts and in the case of capital gains and income taxes, the timing of expenditures as well. Other forms of forest taxes are either partially or totally independant of the timing of incomes and expenditures. A case in point is a version of what foresters call the "ad valorem" tax. Due to confusion with other taxes of the same name, this tax will be referred to as an "accrued income" tax, owing the term to Chisholm (1975). This tax is levied on each increment in forest volume or forest growth. Where (R_4) is the tax rate, the capitalized value of the tax in one rotation is: $R_{4_{t_{m0}}} P\int_0^{t_m} \partial V/\partial t e^{-it} dt$. Noting that tax remittance is made on timber during its course to maturity, the tax is similar to land taxes.

Annual land taxes of a given amount (R_5) are also examined. The land tax is considered along the lines of three alternative definitions. The definitions are:

a. the land tax is a flat yearly per acre liability independent of the biological productivity or economic value of the land;

b. the annual land tax is based on the per acre biological productivity of the land (site index);

c. the annual per acre land tax is based on the perfectly assessed bare land value.

The Excise or Yield Tax.

Where the hauling and harvesting costs are the liability of the buyer, the excise tax falls on the net stumpage (sale) value of the homogeneous stand. The firm's objective function with (R_1), a percentage liability of the final stand value, appears as:

$$(31)\quad W = \frac{(1-R_1)\,PV - wXe^{it}}{e^{it} - 1} \qquad \text{(other notation as before)}$$

Again, the necessary conditions for an unconstrained maximization are:

$$(33a) \qquad (1-R_1)\ P(\frac{\partial V}{\partial X}) - we^{it} = 0$$

$$(33b) \qquad (1-R_1)\ P(\frac{\partial V}{\partial t}) - i\ (\frac{(1-R_1)\ PV - wX}{1-e^{-it}}) = 0$$

The level of the tax affects both equalities in the necessary conditions for a maximum. Changing the magnitude of R_1 will affect both the intensity of establishment and the time length of growth. The excise tax is non neutral.

Since higher levels of the excise tax produce the mathematical equivalence of lower stumpage prices in terms of the firm's investment and output decisions, the impact on per acre supply for changes in the excise tax rate are straight forward. In the substitutes case:

$$(34) \qquad \frac{\partial X}{\partial R_1} < 0\ , \quad \frac{\partial t}{\partial R_1} > 0\ , \quad \frac{\partial (\frac{V}{t})}{\partial R_1} < 0$$

Complementarity between establishment and growth duration produce the following:

$$(35) \qquad \frac{\partial X}{\partial R_1} < 0\ , \quad \frac{\partial t}{\partial R_1} > 0\ , \quad \frac{\partial (\frac{V}{t})}{\partial R_1} < 0$$

Two conclusions regarding the effect of a yield tax levied on the individual firm are warranted. A marginal change in the excise or yield tax from zero to some positive level both reduces the establishment intensity and the firm's supply where the market parameters are constant. If time and stocking are substitutes, the tax will lengthen the optimum rotation age while the inverse result occurs in the case of complementarity. The yield tax is non neutral regarding the optimal mix of (t) and (X) in the individual firm.

<u>The Capital Gains Tax</u>.

Turning the focus to the capital gains tax (R_2), recall that it is defined as a percentage liability of net periodic income. With two decisions (X,t) and instantaneous regeneration, taxable income is the stumpage value net of the regeneration expenses.

Formulation of the firm's objective function appears in equation (36). The analysis is conducted under conditions of a homogeneous product and where the liability of hauling and harvesting is that of the buyer.

$$(36) \quad W = \frac{(1-R_2)\ PV - wXe^{it} + R_2wX}{e^{it} - 1}$$

The first order conditions for the maximization of wealth indicated in equations (33a-b) make clear the importance of a variable level of the establishment investment. If the level of estab-

$$(36a) \quad (1-R_2)\ (\ P(\frac{\partial V}{\partial X}) \quad - we^{it}) \quad + R_2w = 0$$

$$(36b) \quad (1-R_2)\ (\ P(\frac{\partial V}{\partial t}) \quad - \frac{i(PV - wX)}{1-e^{-it}}\ ) = 0$$

lishment is held constant a capital gain tax is neutral. <u>Holding (36a) constant</u>, a change in (R_3) maintains the equality between intertemporal marginal revenue and intertemporal marginal cost. Allowing the optimal level of (X) to vary will result in a tax induced change in both the optimal time of maturity and management intensity.
In the substitutes case:

$$(37) \quad \frac{\partial X}{\partial R_2} < 0\ , \quad \frac{\partial t}{\partial R_2} > 0\ , \quad \frac{\partial\ (\frac{V}{t})}{\partial\ R_2} < 0$$

A marginal change in the rate of income tax (R_2) affects the optimum length of maturity because it affects the equality in equation (33a). With complementarity between the two variables, the following partial derivatives obtain.

$$(38) \quad \frac{\partial X}{\partial R_2} < 0 , \quad \frac{\partial t}{\partial R_2} < 0 , \quad \frac{\partial (\frac{V}{t})}{\partial R_2} < 0$$

In summary, the capital gains tax operates similarly to the excise tax on the individual firm's management plan. Both the establishment intensity and the firm's supply function are reduced with a marginal increase in the capital gains tax, holding market parameters constant. The direction of change of the optimum duration of growth depends on the assumptions about the production function.

The Ordinary Income Tax

Recalling that the ordinary income tax defines the tax base as income net of expenditures, wealth is defined with the income tax included as below.

$$(39) \qquad W = \frac{(1-R_3)\ (PV-wXe^{it})}{e^{it}-1}$$

The first order conditions show that the simultaneous solution of (X) and (t) are independant of the level of income tax. The income tax is neutral within the firm providing of course that the tax doesn't alter land allocation.

$$(39a) \qquad (1-R_3)[(P\partial V/\partial X) - we^{it}] = 0$$

$$(39b) \qquad (1-R_3)[(P\partial V/\partial t) - i(\frac{PV-wX}{1-e^{-it}})] = 0$$

The Accrued Income Tax

Consider now the accrued income tax. Formulation of the present value of the net of tax income stream is shown in equation (40).[2]

$$(40) \qquad W = \frac{PV - R_4\int_0^{t_m} P(\partial V/\partial t)e^{it}dt \ - wXe^{it}}{e^{it} -1}$$

The first order conditions indicate that the optimum level of establishment and maturity depend upon the magnitude of the tax rate.

(40a) $P(\partial V/\partial X) - R_4 \int P(\partial^2 V/\partial t\, \partial X)e^{it}dt - we^{it} = 0$

(40b) $P(\partial V/\partial t) - i\left[\frac{PV - R_4 P\int(\partial V/\partial t)e^{it}dt - wX}{1-e^{-it}}\right]$

$- R_4(P(\partial V/\partial t))e^{it} = 0$

The necessary conditions for a maximum show that the optimum level of establishment and maturity will depend on the level of the accrued income tax. A parametric shift in the level of the tax will produce shifts in supply similar to those outlined in Chapter 2. Whether the time of maturity increases or decreases will depend on the relationship between the variables in the production function, however.

(41) $\partial X/\partial R_4 < 0,\ \partial(V/t)/\partial R_3 < 0,$

Taxes on Land

Three kinds of annual per acre land taxes are investigated in this section. They are: a flat dollar amount on all private timber land; a dollar amount which increases with site productivity or site classes for private timber land; and a percentage of the perfectly assessed annual site rent or bare land value for all private timber land. Each of these taxes operate similarly within the firm. Marginal changes in these land taxes are neutral in terms of the firm's management plan providing they do not produce negative timber land rents This point can be demonstrated by viewing the firm's objective function with a yearly land tax (R_5) in equation (42). It is necessary to use a slightly different annual interest rate equivalent (i_Y) to compound the annual land tax because of the use of continuous time in the forest growth and financial valuation portion of the equation.

(42) $$W = \frac{PV - wXe^{it} - R_5(i_Y)^{-1}(e^{it}-1)}{e^{it} - 1}$$

The first order conditions indicate that (X) and (t) are independent of the level of the land tax.

(42a) $P(\partial V/\partial X) - we^{-it} = 0$

(42b) $P(\partial V/\partial t) - i\left(\frac{PV-wX}{1-e^{-it}}\right) = 0$

The impact of a special land tax falls only on the wealth of the timber firm. The tax reduces the value of the land input when employed by the timber firm. Within the firm, the land taxes are intertemporally neutral regarding the allocation of scarce productive factors. A marginal increase in (R_5) will marginally reduce land rent in timber use regardless of whether the tax is based on biological productivity, land value, or simply a flat per acre amount.

THE INCIDENCE OF FOREST TAXES

Turning to the larger issue of the incidence of the forest taxes in the timber market, the pretax equilibrium in Figure 4.1 is at output Q_e and price P_e. Total supply is the sum of the two sector contributions as was mentioned in the previous chapter.

Acknowledging the discussion in the last chapter regarding the relationship between the individual firm's supply function and private sector supply, excise taxes, capital gains taxes, and accrued income taxes produce the following market results. Because these taxes are non neutral within the firm, increases in any of these taxes will shift private timber supplies to the left holding the quantity of land constant. In addition, because these taxes reduce rent, land use will change from timber production. These two changes indicate that the private sector will produce less output at the original price. The combined effect of changes in land use and the firm's optimum management regime is to shift (reduce) private sector supply to S_1' in Figure 4.1. These tax induced changes in land use and the firms' management regimes will continue until the reduction in

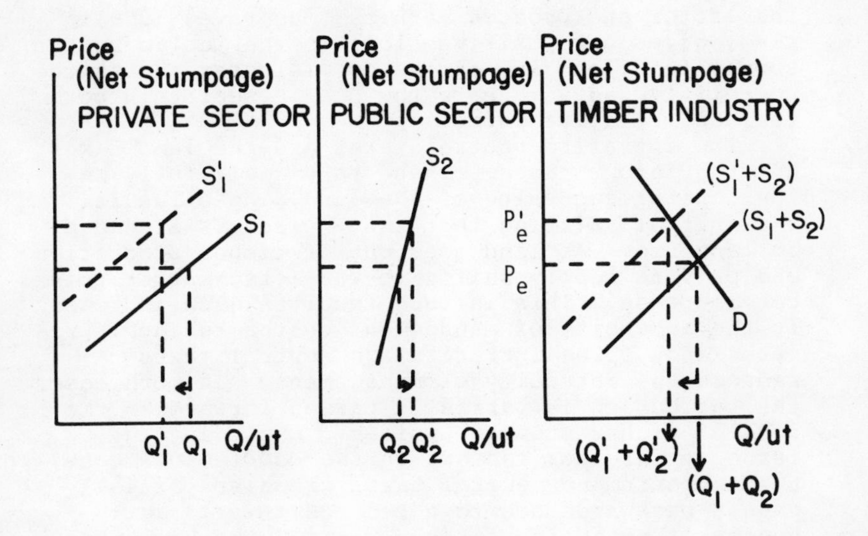

Figure 4.1

supply is sufficiently large to produce a higher timber price which again will equilibrate the market in terms of land use and efficiency in the firms' investments. The shift in private sector supply produces a corresponding shift in aggregate market supply from $(S_1 + S_2)$ to $(S_1' + S_2)$. At the new post tax supply change, market clearing price becomes P_e' with the corresponding output $Q_1' + Q_2'$. The marginal land or firm will again be indifferent between timber production and other land uses and conditions of efficiency will again hold in establishment and the time length of growth.

Holding market demand and the public supply function constant, an increase in the intrafirm non neutral forest taxes reduces timber supply and increases timber price. There is less wood available and consumers pay more for it. The burden of the tax is partially passed forward to consumers of final wood products. Additionally, the burden of the tax is partially passed backwards to suppliers of establishment factors (ignoring the question of inferior establishment factors). The extent to which the tax burden is shifted depends upon the elasticities of supply and demand in both the factor and product markets (Musgrave, 1959). Also, given any positive slope to the public sector supply function, the excise or yield tax shifts the competitive advantage to the public sector in both real and relative terms.

The intrafirm neutral taxes are the land taxes and the income taxes. These taxes fall fully on the capitalized value of the land. As a result, the initial shifting that takes place is a change in land use. As land goes out of timber production and private supply shifts to the left, market prices increase. This in turn results in an increase in the intensity of management on the residual private lands. The intrafirm non neutral taxes reduced the intensity of management. In both cases the tax burden is partially passed forward in the form of higher stumpage prices. Excluding the issue of inferior factors in the complements case, the intrafirm nonneutral taxes are also partially passed backwards due to a reduced rate of frequency of establishment purchases. The backward passing of the tax burden is ambiguous where the taxes are intra firm neutral. It depends upon whether the greater intensity of management on the residual private land base more than offsets the tax induced reduction in the land base.

The importance in distinguishing between the different kinds of land taxes lies in the different patterns produced in changing land uses. Recalling that marginal land is that land whose timber value is equal to its alternative use value, only the flat per acre tax will shift land at the margin. Because both location and biological factors affect the value of land in timber as well as in other uses, the tax which is independent of these factors will insure that land closest in value to its next highest use will be shifted out of timber production. Both the assessed value tax and the site index tax will shift more non marginal land out of timber use than the flat per acre land tax. The flat land tax rate would have to fall between the high-low extremes of the assessed value tax and site index tax in order to produce equivalent tax revenue before shifting took place. There is no reason to believe that the most biological productive nor the highest economically valued timber

Figure 4.2a

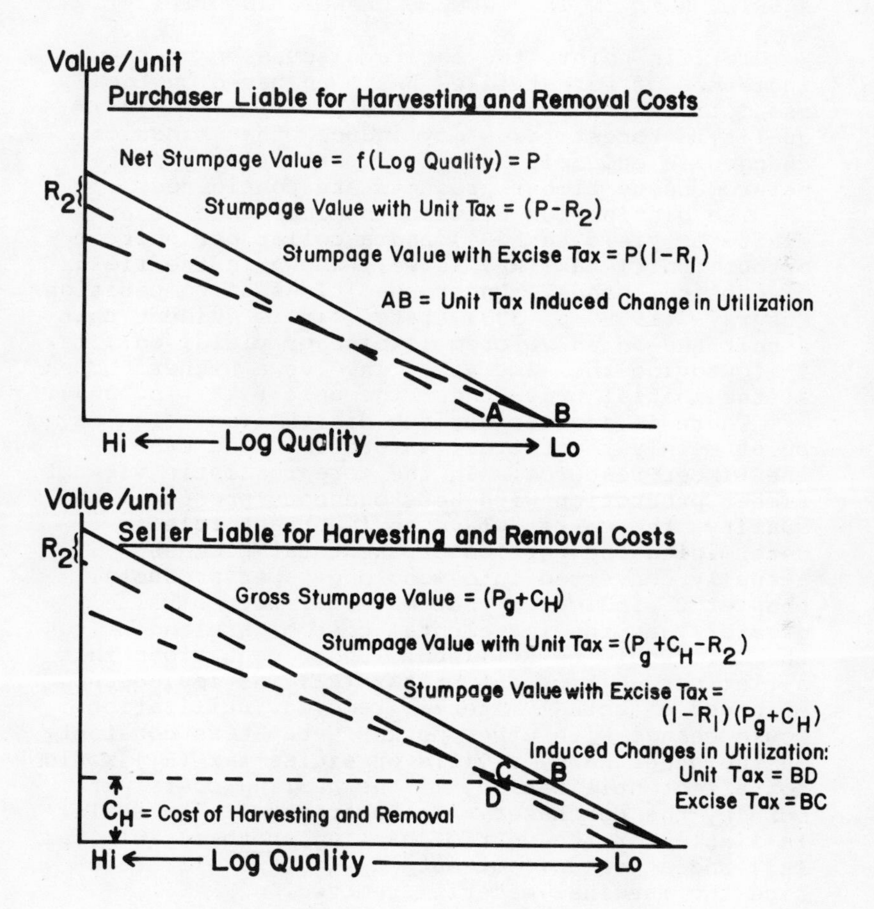

Figure 4.2b

land is necessarily the most valued in other uses. As a result, both the site index tax and the assessed value tax will both shift some land that is not as close to the margin as would be the case of a flat land tax.

A BRIEF NOTE ON TAX INDUCED CHANGES IN UTILIZATION

To this point, the entire discussion of the incidence of forest taxes has been based on the assumption of homogeneous forest stands and fibre quality. Forest taxes may induce other kinds of changes in economic behavior, particularly where heterogeneous timber products are considered.

The distinction between the incidence of an excise or yield tax (R_1) and a dollar per unit of output (unit) tax (R_6) is well known in the field of public finance. Under conditions of competition Musgrave (1959, p. 303) states "it is evident that a unit and an ad valorem (excise or yield, ed.) tax that provide the same yield involve a higher burden at the initial price under the unit tax" (p. 303).

There is another curious distinction which may be of mainly theoretical interest in the case of the timber resource. In the more realistic view of timber production with heterogeneous product quality, the market plays an important role in determining the portion of the total biomass actually converted into wood and paper products. Chapter 2 indicated that material with a value greater than the incremental cost of hauling it to the mill would be utilized. Under conditions that a flat per unit of output tax (R_6) was levied and enforced on actual material removed, utilization would change with other market parameters constant. On the other hand, a yield or excise tax (R_1) would not affect utilization, if the hauling costs were born by the purchaser. In contrast, if the seller is liable for the cost of hauling, both a yield tax (R_1) and a per unit of output tax (R_6) will redefine the marginal material removed.

Figures 4.2a-b should help clarify these points. In diagram 4.2a, where the purchaser is liable for hauling, the revenue function with the yield tax (R_1) intersects zero at the same point as the revenue function without the tax. On the other hand, the revenue function with the per unit tax (R_6) intersects zero to the left in Figure 4.2a, indicating that the tax reduces the merchantable portion of the total harvest. With constant market parameters and the liability for hauling on the shoulders of the purchaser, only the excise tax (R_1) is neutral with respect to utilization.

Figure 4.2b represents the effect of the two taxes under conditions where the liability of the expense for hauling is the responsibility of the producer-seller. The sale price is then the gross stumpage value. In this instance, both the excise tax (R_1) and the per unit output tax (R_6) affect stand utilization by redefining the marginal merchantable unit.

In the larger context of the market, changes in stand utilization along with changes in the management regime will not only affect total quantity offered for sale, but also will affect the relative prices of the various grades of logs or other harvested material.

SOME EARLIER TAX FINDINGS

As in the rediscovery of financial maturity, Gaffney's (1976) article provides a benchmark for forest tax analysis. Yet, unlike his work on financial maturity, Gaffney's early tax foray is often ambiguous regarding the conditions for which the results hold, as well as the rationale for choosing the rate of turnover of capital as a welfare criteria. Thomson and Goldstein (1971) vigorously attack the findings of Gaffney. Yet they too are careless in terms of stating the conditions for which their results hold and in the process of relaxing an assumption, make a mistake of their own. Chisholm (1975) finally provides some clarity to the differences between Gaffney and Thomson and Goldstein. Gaffney is investigating economy wide tax policies while Thomson and Goldstein approach the problem as was done here with the rate of interest constant.

Gaffney is concerned with taxes that slow the rate of turnover of capital. Yet he never defines capital. Since Gaffney assumes that the value of land is zero, timber must be the only relevant capital. Taxes that lengthen the harvest date will automatically increase the rate of turnover of timber in the ages relevant to maturity (holding X constant). Under the conditions assumed by Gaffney, the turnover of timber capital is virtually useless as a social welfare norm.

All three articles really ignore the intensity of timber management and its effect on supply. Thomson and Goldstein recognize in a footnote that planting may either be a complement or substitute for the time length of growth. Unfortunately, they fail to carry the point further stating: "To finesse these complexities, we treat the planting cost as constant, a procedure the literature generally sanctions." From this perspective, Thomson and Goldstein determine which taxes are neutral on the basis of financial maturity alone. Specifically, they determine that the capital gains tax is neutral. It was shown here that the capital gains tax is nonneutral because it affected establishment intensity which in turn affected maturity when the necessary conditions are simultaneously solved. When Thomson and Goldstein move from the analysis of tax neutrality to the subject of shifting the tax burden, they relax the assumption that maturity is the only decision variable. In the process of relaxing that assumption, they specifically fail to recognize that the capital gains tax is nonneutral.

Chisholm looks at shifting the tax burden in terms of changing land uses and values. While his reasoning is clear, he avoids the most interesting question regarding tax induced changes in timber supply, namely how will changes in rotation affect the supply and price of stumpage?

All three articles lack a positive theory of timber supply as a logical basis for an extended investigation of the incidence of forest taxes. As a result, the welfare implications of tax mechanisms are wholly inadequate. A tax which increases the optimum harvest age along a fixed yield function will normally be associated with a higher mean annual increment. Holding demand constant, all three articles have found several forest taxes which increase supply and reduce timber prices while raising revenues for presumably valuable government goods and services in the absence of shifting land use. At face value, these findings should go down in the annals of economic thought, yet none of the authors mention the possibility of such occurances nor do they prove that their shifting mechanisms will offset the otherwise positive supply responses.

FOREST TAXES SUMMARIZED

The major portion of the forest taxation study

presented in this chapter incorporates the assumptions of homogeneous output and two production decision variables. In the light of those assumptions, it was found that the sales or excise tax (R_1), the capital gains tax (R_2), and the accrued income tax (R_4) were nonneutral within the firm's management plan. As a result, industry supply shifted with the imposition of either of these three taxes, thereby shifting a portion of the tax burden to input suppliers or output purchasers. Additionally, to the extent that rent in timber production on private marginal land is reduced, higher excise or income taxes on timber land will shift land use to non timber alternatives.

Three kinds of land taxes and the ordinary income tax were also discussed. While each of these are neutral, they were all found to be nonneutral under the conditions that timber land use is granted special taxation consideration and resulted in different patterns of shifting land usage. Special taxation status of private forest land regarding land taxes implies that the initial tax impact falls completely upon rent in timber land use. If timber land taxes are lower (higher) than on other land uses, marginal land will be allocated toward (away from) the timber production use. It was also noted that where timber land is granted special taxation status, a flat per acre land tax is more efficient in determining land use than are land taxes based either on site classes or on assessed value.

The issue of mixed public-private timber land ownership raises some troublesome taxation problems. Any tax levied only on a portion of the total land in existence (private timber land) shifts the competitive advantage to the non taxed public timber sector of the industry. Assuming that the public timber supply function is unaffected by the institution of forest taxes on the private secotr, only a portion of the industry faces the higher operating costs resulting from the taxes. For a given level of imposed tax rates, the stationery public supply function implies that private supply will shift more to mitigate a portion of the tax burden than would be the case if the tax rates were to be applied to the industry as a whole.

Tax induced changes in the rate of turnover of land are not very important findings by themselves.

Tax induced changes in the supply of timber is the more important welfare consideration due to its relationship to the availability and price of final wood products. Holding market parameters and the supply of timber land constant, the direction of movement of a tax induced change in optimum rotation age depends upon the character of time in the production process. Taxes may lengthen or shorten the optimum growth period, but so may a change in the product price. A longer or shorter time length of growth does not, by itself, indicate any change in economic efficiency.

Given the information on shifting the burden of forest taxes, it is clear that private producers would prefer either the excise tax or the capital gains tax over the land taxes discussed. At constant market parameters and holding tax yield to the public revenues equal, private producers can shift a portion of the burden both forward and backward under the excise or income taxes by changing their management regimes. This is not the case under the same conditions for the other taxes. The only potential for shifting the burden of special forest land taxes is by changing land use and thereby timber supply. Unfortunately, there does not appear to be a sound method to compare the magnitude of output changes with respect to marginal changes in either the income or excise taxes. If this were not the case, propositions could be presented regarding producers preferences for either the income or excise tax.

Interpretation of the results presented on tax induced changes in utilization is somewhat cloudy. The analysis was presented under conditions in which the taxes were levied and enforced on the actual material removed. Where there are heterogeneous timber products, the tax would have to be levied on the basis of either total volume removed (unit tax) or on the basis of the total volume and value of each grade removed (excise tax). Clearly, in order for either of these taxes to actually induce changes in utilization, actual measurement (scaling) of merchantable volume removed is necessary. If either of these two taxes were levied on estimates of volume at harvest from some standard yield table, or were levied on reported harvest volumes, their impact on stand utilization is not as straight forward. Yet the relationship between tax induced changes in timber utilization and the broader area of efficient recovery of stock natural resources such as coal or petroleum should be made (See for example, Scott, (1953), pp. 184-190). At the time of harvest, the forest stand is a stock resource to be utilized efficiently. As is often the case of other natural resources, recovery costs are not constant over either a range of resource qualities or extraction methods.

In conclusion, it is well to point out that the results of much of this study differ markedly from the Gaffney and the Thomson and Goldstein studies. Part of this contrast must be attributed to the fact that different assumptions were used in the respective analyses. It is believed, however, that the most fruitful manner to investigate the incidence of forest taxes is to place the principal analytical focus on tax induced changes in production and supply. The theory of timber supply is necessary in order to assure that the tax is first imposed on an initial stationary state, and that the tax disturbance will result in adjustment to a subsequent stationary state.

NOTES

1. There is a form of revenue sharing with either a portion of the federal timber sales receipts being turned over to the counties, or in lieu of tax payments are made on the basis of national forest acreage as well as other criteria.

2. Thanks go to Howard Rhinehart for assistance with the formulation of this forest tax.

5 Research and Policy Implications

INTRODUCTION

Rather than reiterate the findings of the previous chapters in an overall summary, the concluding chapter attempts to develop a larger perspective for the results. By outlining some areas of future potential research interest, and pointing out others of operational significance to timber managers and interested practitioners, the scope of the study can be seen in a different light.

IMPLICATIONS FOR OPERATIONAL TIMBER MANAGEMENT

Forestry economics differs from theoretical economics due to its explicitly applied nature. It is important in forestry economics to relate theoretical findings in a manner which is of potential usefulness to decision makers. In theoretical economics, marginal analysis often becomes almost more than second nature to the practitioner, occasionally to the extent that the plight of the managerial decision maker is overlooked. Theoretical economics relies heavily on the assumption of continuous functions, as was done in this study, in the development of testable propositions. Unfortunately, continuous functions and their analytical luxury are not always an every day fact of contemporary business life.

Since World War II, a whole set of optimization techniques has evolved that are more amenable to making operational decisions in the light of the realities of business data and information. These techniques, which are often called programming, or operations research techniques, do not normally require continuous functions for solving numerical economic problems.

Dorfman (1953) summarizes these ideas nicely with the following statement:

> The central formal problem of economics is the problem of allocating scarce resources so as to maximize some predetermined objective. The standard formulation of this problem--the so called marginal analysis--has led to conclusions of many questions of social and economic policy. But it is a fact of common knowledge that this mode of analysis has not recommended itself to men of affairs for the practical solution to their economic and business problems. (p. 797)

In terms of consistency between marginal solutions and programming solutions, Dorfman states "that the equilibrium position of a perfectly competitive economy is the same as the optimal solution to the mathematical (programming, ed.) problem embodying the same data." (p. 824) This is the result of the fact that the optimization techniques were developed in order to solve real problems which previously only had theoretical solutions.

Many of the theoretical findings presented in this research have practical importance in light of solving real management problems. The theoretical findings regarding the timing of investment decisions is an important case in point. Previous attempts to develop "optimum" management regimes either completely ignored the question of timing of intermediate stand life management decisions, or with no marginal solution in mind, tried all possible combinations of timings and intensities in order to arrive at a solution. The theory of timing presented here can assist in short cutting these latter kinds of computer solutions. The theory specifies a set of conditions which an optimum solution must achieve. Use of these conditions as constraints in solving practical problems may only lower computer expenses, or they may allow a wider choice of routines to solve problems (e.g., linear or nonlinear programming versus dynamic programming).

Of at least equal importance to methods of solving problems is the relevance of the conditions for optimal timing to the interpretation of results. The manager can, for example, explain why the optimum age for thinning is thirty-seven years as opposed to some other time. These kinds of advantages should not be overlooked in classroom teaching as well. The presumption here is that the theory of timing intermediate inputs coupled with methods of solving managerial problems is a preferable classroom experience to either by itself.

THE TREATMENT OF RISK

Risk, or probabilistic occurrence of events, is of great importance both in forest production and forest products markets. Risk was ignored in this study in order to come to grips with the problem of integrated investment and supply. An example of an earlier and more thorough examination of risk is Dowdle's "Investment Theory and Forest Management Planning" (1962). Dowdle used the expected returns -variance of returns rule developed by Markowitz in analyzing forest investments with risk. Other contributions such as Burt's (1965) don't appear to have received the attention that is warranted.

The treatment of market speculation ignored the economic difference between the cognitive recognition of short term changes in price as opposed to "once and for all changes in demand." The analysis only presumed that producers were able to distinguish between these situations. Ignoring the behavioral aspects of the formulation of expectations is only partially justifiable in the context of the market. It would be far more appealing to indicate theoretically how these expectations are formed. Certainly, changes in the mortgage rate relative to some other interest rate index may be a short term phenomenon. Yet is is presumed that the producer has the capacity to distinguish this change along with other simultaneous changes in demand and supply information. In order for someone to make a long term commitment to timber production, some idea of the profitability of the decision relative to other alternatives must be known. As timber output is decreasingly composed of old growth timber, further research into the formation of expectations and the attendant investment behavior becomes increasingly important.

SUPPLY IMPLICATIONS AND THE FUTURE

Perhaps the most significant finding in terms of professional foresters is one that many economists would take for granted. It has been proven on the basis of some elementary biological relationships that the equilibrium private supply is a positive function of the market clearing price. There was no need to introduce technological changes into the model in order to derive a stable market supply function. Any government program designed to subsidize timber production in order to maintain a low price will inhibit the incentive of private producers to invest and increase output. In fact, holding technology, factor costs and the interest rate constant, the major means of increasing private supply is through in timber demand.

The only theoretical question of empirical importance is the issue of continuous private supply over time with the decision variables of time and establishment as complements. This issue may be analogous to the question of stability with only two complementary factors in a static production function. In that case, firm size is not stable. In this case, firm size is stable, but the adjustment to a longer rotation may not be stable in the market context. This finding is likely an interesting theoretical result without much real world significance. Producers are faced with many more than two decision variables. However, it does appear that some additional research attention should focus on adjustments where time is a net complement to other productive factors in the management regime.

In general terms, it was found that the derivation of a timber supply function was analogous to the theory of supply under conditions of a static production function. The main differences were that quantity at time of harvest varied with different price levels as well as the frequency of harvests. This finding has a theoretical predecessor in Alchian's (1959) important contribution entitled, "Costs and Outputs". It is perhaps, on the other hand, more commpletely stated than Alchian's article in terms of timber and a production function over time.

THE SITE: A FIXED OR VARIABLE FACTOR OF PRODUCTION

Throughout the study, the biological productivity of the site was constant and there were constant returns to scale in terms of the size of the ownership. Management activity in one stand did not affect the productivity of the site in future stand communities. This is the typical approach to the infinite series of regenerations but it does not necessarily mesh completely with biological reality. For example, some harvesting techniques may compact the soil more than others, resulting in a lower biological capacity of the future site. At the same time, the effect of fertilizing current stands on future stand communities is not clearly understood. Possibly, some of the added nutrients become a part of the stock in the site and hence improve the growth possibilities of future untreated sands. In other words, management activity may subtly or even more decidedly affect the biological potential for future production. The production function in future stands may be a function of the current management regime.

It is technologically feasible, if not economically efficient, to more directly alter the productivity of forest sites. Terracing, irrigation, and weather modification are all means of changing the production per acre, or better yet, the site's productive potential. In the main, the implementation of these activities will depend on current and future timber prices along with production costs.

For the investment analyst, a variable site productivity makes determination of the optimum regime more difficult but not impossible. A plan for an infinite series of harvests would be a long run plan involving a series of different production decisions in each rotation. It could be that there would evolve an optimum rate of depletion of the site where management activity erodes productivity. The game under these conditions may be intellectual rather than real, but it would imply the determination of the optimum rate of depletion of a strategic factor in producting a renewable resource. If the management activity improved site productivity over time, the question would be more familiar in terms of the optimum rate of investment.

Returns to scale in terms of the size of ownership and individual stand size are typically ignored in viewing per acre yields and costs. In order to come to grips with a better descriptive theory of the firm, the quantity of land should be variable as well. It appears as though a feasible manner of handling the problem of optimum firm size lies along the lines of location theory. As land is added to the firm, it will be generally of inferior location regarding both stand management costs and hauling costs via proximity to mills. Dealing with the firm in this manner implies that per acre costs of treatment increase due to larger area size and distance and per acre net stumpage values decrease due to greater hauling distance and costs. In a more refined theory of the firm, such as this, the physical per acre production opportunity set is not a function of the total land holdings, but factor costs such as those for stand establishment along with net stumpage value are. Stand management would vary for different sites in the firm's portfolio depending on distance and the associated cost of operation, site productivity, and other market detemined parameters. It is believed that the inclusion of land ownership size would embellish the results of this study but not substantially alter them.

FOREST TAXATION

Due to the nature of mixed public-private production and the legal constraints on taxing public output, a review of the entire taxation system and revenue sharing of federal states and local governments is in order. It may make a good deal more sense, for example, to tax purchasers of timber rather than producers in the light of legal constraints. Revenue raised from a tax on purchasers would have to be in lieu of the current revenue sharing arrangements from federal timber sales to states and counties. The tax policy would have to be written in such a way that a major portion of the burden would be shifted back to the producer in the form of lower stumpage prices. Wealth or land rents would be reduced via lower stumpage prices and there would be no bias in favor of public production.

Further analysis of different kinds of taxes on purchasers is needed. Certainly, the integrated forest products firms can be characterized by "internal purchases" at best. Given these ramifications, it might, for example, be desirable to define the harvester as the purchaser and levy harvesting license fees. The license cost of operating in a particular locale would fall heavily on the bid price for any market transactions and on the wealth of the integrated firm.

The administrative desirability of various kinds of forest taxes is a relatively unexplored subject area. Many issues of importance come to mind. Whether or not forest tax laws are easily enforced and/or induce cheating (thereby overtaxing the honest taxpayer) is a question of administrative equity. For example, a yield tax or a per unit of output tax requires knowledge of either the total revenue or total volume of the harvest. In the case of the integrated firm which grows its own timber, no market transaction is available for tax audits. Liability may be derived from self reported estimated volumes (Washington State) and values must be derived in the absence of a direct sale. No doubt, self reporting systems are desirable from the producers point of view, but the cost of measuring output or even spot checking self reported volumes is quite high. The concern must lie with whether or not the value of accurate stumpage information exceeds the cost of gathering the information.

Income taxes and capital gains taxes present different kinds of administrative problems. Lands forming the tax base in states or counties may be owned by firms located outside the taxation district. Not only would it be difficult to allocate a portion of the firm's taxable income to the particular tax locale, but if the state were the taxing unit, methods of allocating the revenue raised to the various counties in lieu of the property tax would also have to be forthcoming.

A property tax that is independent of assessed value would certainly be the easiest to administer in terms of determining tax liability and enforcement. This form of tax does not appear to be without its administrative problems, however. If the tax was levied by the state, and revenues were to be allocated to counties, another allocation problem arises. Should the allocation be based on land area or demand for public services? Those counties with the greatest population densities tend to require the greatest revenues for public services. Counties with low population densities and large land areas would likely resent "subsidizing" the counties with large populations.

Property taxes based on assessed values directly imply administrative costs of assessment. In addition, the assessment function is likely to be somewhat arbitrary in nature (e.g., variation in site productivity may require intensive investigation). If assessment were perfect, all land were competitively owned, and the assessment rate was a constant percentage of assessed value, allocation of land to various uses would not change. We can only wonder what the administrative costs of perfect assessment would be. To the extent that assessment is imperfect, the assessment system of taxation is inequitable. Taxpayers are not treated as equals under the imperfect assessment system. Wealth of some taxpayers escapes taxation more than the wealth of others. Also under the assessment system, revenues produced in some areas may again be inadequate to meet varying demand for local public services.

Taxation has been seen to induce changes in patterns of land ownership and use, both of which appear to be of increasing interest to politicians and policy makers. This study implies that differential land taxes and income taxes are a more efficient tool for intentionally changing patterns of land use than are the other taxes examined. Any differential treatment of forest lands in terms of the property or income tax affects the incentive to own land directly. Other non neutral taxes affect both the production decision variables and the incentive to allocate land to timber production. It should be noted that preferential treatment of timber land may lead to shirking of taxes in another way. It is easy to imagine, for example, an owner classifying his land as forest land at some preferential rate until he is ready to sell it for recreational development which might be liable for a higher tax rate. Only the owner would be able to testify about the true nature of his intentions at the time of the original tax classification.

Land taxes should not be construed as being the only tax tool for solving environmental problems that some or all people don't want to tolerate. Land use zoning may be called upon to assist in solving problems of non market costs or benefits and/or common property problems. Alternatively, in the case of the emission of non market effluents, a per unit tax on effluents emitted may be the best. In the area of restricting over utilization of common properties, a license or head tax on users may serve to assist in achieving the socially desired level of utilization.

While many forestry professionals may believe that timber production should be a special case for comparative tax advantage, the grounds for such a case appear to be limited. The competitive advantage associated with non-taxed public lands may warrant special consideration for private producers. If this problem can be alleviated via taxation of harvesters as was previously mentioned, higher economic efficiency can be achieved and the only possible further case for special tax consideration would have to rest on some non market explanation. There appears to be no special market grounds for private timber production as a particularly "needy" industry beyond the non-tax status of the public sector. The market appears to assure a timber supply as long as demand conditions continue to warrant it.

RESEARCH AND PUBLIC TIMBER PRODUCTION

The major emphasis of this study is on a theory of market determined private investment and supply. A purely competitive model of the private sector was developed to indicate the efficient means of producing timber. As was previously indicated, an incontrovertible necessary condition for the maximization of social welfare is the condition of efficiency of production.

In analyzing the economic implications for public policies, it is often desirable to utilize the competitive norm as a standard of comparison. Often the competitive solution is the important norm for comparison with the _modus operandi_. If the findings concerning efficiency in timber production that have been developed in this study are true, then the policy implications for current public sustained yield management are dramatic. The current sustained yield management approach can not be substantiated on the grounds of economic efficiency. However, there may be other aspects of this practice that warrant examination before the current sustained yield approach is completely rejected. For example, the remote possibility exists that public timber production in its current subsidized manner might be justifiable in terms of consumers surplus added to purchasers of final products. Since there is no adequate proof of the existence of this contradictory welfare phenomena, it must remain only hypothetical.

Yet another facet of public timber management warrants close policy scrutiny. The terms of the public timber sale contract, as is the case of all sales contracts, either specify or imply the dispersion of risk between competing parties. In addition, the contract determines the actual rights to use that are exchanged, along with when and how they shall be exchanged.

In order to illustrate the importance of these points, two aspects of National Forest Service timber sales contracts are cited. It is important to point out that these sales are typically "scaled" (payment obligation based on measured volume of timber actually removed times the price er unit of volume).

General Forest Service policy (Forest Service Manual) suggests that the period of the contract shall depend upon the total volume sold as follows:

Volume (V) in m.m.b.f.	# Operating Seasons
$V \leq 2$	2
$2 < V \leq 5$	3
$5 < V \leq 10$	4
$10 < V \leq 25$	5
$25 < V$	6 or more

In order for the buyer to fulfill his contractual obligations, he must harvest and remove his timber within the specified period of the contract (contract periods may be renegotiated under stated limited conditions). It is notable that the length of time increases less than proportionally with the sales volume. Even with scale economies, it may well be that a buyer cannot afford to temporarily restrict harvesting for a turn around in market conditions. In other words, the period of the contract places constraints on the buyer's incentive to adjust his rate of output to meet changing market conditions.

Other important aspects of many sales contracts are clauses that call for "escalation", "rate redetermination", and "scheduled rate redetermination". Those contracts that specify scheduled rate redetermination indicate that at specific date(s) after the contract commences, all or a portion of the timber still unharvested will be reappraised in light of the then current market conditions. The alternative escalation clause in the contract is defined in the <u>Forest Service Manual</u> in the following manner.

> ...where the quarter index average is below the base index, adjust bid rate downward by full amount (100%) of such difference, but not below base rate. When quarterly index is above the base index, the bid rate is adjusted upward by one-half (50%) of the difference.

The base rate is specified in the original appraisal and the original bid rate arrived at in the competitive bidding process is "escalated" quarterly by the updated price index. Many contracts may not have either the scheduled rate redetermination clause or the escalation clause, but all have the rate redetermination clause (non scheduled). This latter general clause allows the buyer to request a reappraisal for reasons of changes in market conditions along with other changes.

Under contract terms such as these, the Forest Service or taxpaying public bears a great deal of the burden of risk due to changing market conditions. If these and other clauses in the contract tend to make harvest rates on public land sticky to changing prices, the contract may actually partly cause the dramatic short term changes in the price itself. All of the demand fluctuations are reflected in price changes rather than public output changes. It must be added that public assumption of the market risks likely increase the price that competitive bidders are willing to pay for timber.

There is a recently emerging literature of the economic implications of alternative contractual mechanisms for disposing of timber. Nautiyal and Love (1971) and Bueter and Arney (1972) are early reference points which examined various means of charging for stumpage sold. <u>The Report of the President's Panel on Timber and the Environment</u> (1973) suggested several changes in current National forest timber sales contracts. Jackson (1976) presented a general framework for analyzing the benefits and costs of changing contract clauses. Johnson (1979) has completed the most exhaustive statistical analysis of timber sales contract clauses in a choice theoretic framework.

Waggener's (1969) examination of public timber supply policies which are designed to stabilize the forest products industry reveals important price and output dynamics issues. His basic style of analysis is used in order to reemphasize important intersectoral impact questions. Public policy makers are confronted with two supply policy alternatives on the national forests. Policy alternative one is embodied with supply function S_B on the national lands. This policy is designed to

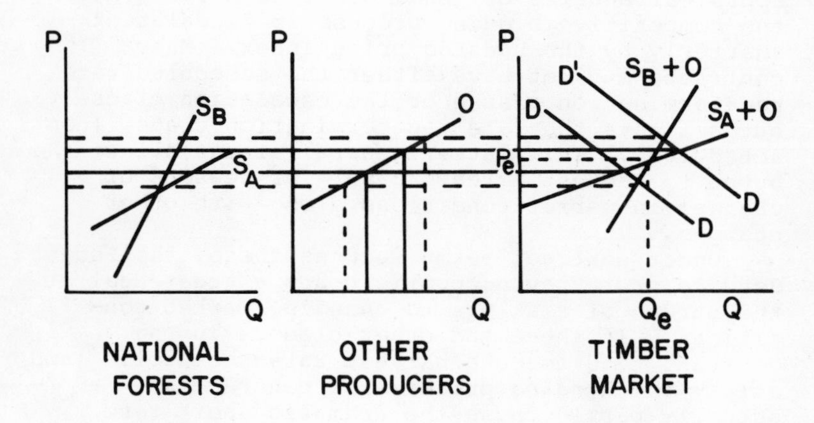

Figure 5.1

minimize the variation in employment over time for those whose employment is directly based upon national forest timber sales. The second public supply alternative is shown as S_A. This price responsive supply policy is basically one in which the national forests would store timber on the stump during low points in the business cycle and then increase the rate of logging during the periods of heavy woods products demand. The middle part of the graph represents the "other" part of the national timber supply scene. Other timber producers are the small woodlots, the large industrial owners, the state and local governments and so forth. Aggregate market supply is shown at the right hand side of the diagram. Note that there are two alternative market supply functions. Supply function (S_B + 0) is the market supply function owing to the goal of supply "stability", and (S_A + 0) is the market supply function owing to a price responsive public supply policy. At price P_e, where the two market supply alternatives intersect, Function (S_A + 0) is more elastic due to the greater slope of S_A than is alternative function (S_B + 0). Output Q_e represents the aggregate anticipated output goals for the suppliers of timber. What is of importance are the responses produced under the two supply alternatives resulting from demand fluctuations around the anticipated market equilibrium (Pe at Qe). Suppose that demand fluctuates between DD and D'D' in Figure 4.1. The tradeoff between output fluctuations and price fluctuations is apparent in the right hand side of Figure 5.1. Under the price responsive alternative, market prices will vary less while market output will vary more. However the differences in fluctuations under the two alternatives is transmitted throughout the industry in different ways. Under the stable public supply alternative, output varies more in the "other" sector than under the price responsive public supply policy. If variation is one of the things that we don't like, then one of the costs of "stability" in the public (national forest) sector is greater output variation elsewhere.

Whether national forest timber supply varies less over time than it would if the forests didn't have the "stability" goal, is an empirical question. Policies are often simply symbols or window dressing. Yet the proposition can be made that if the national forest want to stabilize local communities through special supply policies, then they will destabilize communities elsewhere.

TIMBER PRODUCTION AND UNCOMPENSATED BENEFITS AND COSTS

At the outset, any multiple resource occurances were ruled out via assumption. The merits of this assumption are based on the desire to develop a theory of timber investment and supply under conditions of competition. Forest management would be far simpler, if less rewarding, were the trees the exclusive forest resource. Yet just as forests are is found during fall hunting season. In turn the game population itself may be owned by still other parties. Divided resource tenure such as this can easily lead to inefficient allocation where transaction costs exist. Second, government agencies may choose to underprice resources which are in principle exclusive in nature. Examples are too numerous to mention completely but grazing rights and irrigation rights are two oft cited cases. Elements of public goods may also be found in the area of multiple use. An example might be concurrent consumption of the view or "aesthetics". Lack of adequate value information coupled with joint products production functions adds further misery to those in search of efficiency in resource allocation.

It is certainly appropriate to inquire into the social costs of managing lands for the exclusive production of timber. In the absence of appropriate non market values and joint product production functions, a lower level of choices is still available. What is the opportunity cost of forgone timber value of any action which is seen as protecting or enhancing non market benefits? Admittedly, the unit of exchange in choices of this kind is akin to the "wooden nickel". Yet the least cost solution where costs are determined from an initial "efficient" allocation of a single resource can lead to the following question. If there is more than one way to accomplish a given multiple use objective, which one has the lowest opportunity cost of timber? Alternatively, how much must the multiple use values be worth in order to forgoe known timber value? In both instances, the theory of efficient timber production is of immense value.

ECONOMICS, SILVICULTURE, AND THE NEED FOR ADDED ANALYTICAL PRECISION

Consider now some of the implications of relaxing the initial assumptions of clearcutting, even aged management, and homogeneous wood products. Jackson and McQuillan (1979) have recently predicted the manner by which tree size, logging methods, reproduction systems (i.e., clearcutting and seed tree cutting versus shelterwood cutting and selection cutting) will affect the sale value of a given stand in a given market. This equation for conversion returns makes it possible to complex biological and physical entities, so too are the social values associated with their occurance and manipulation.

From a decision framework there appears to be two distinct approaches to the management of multiple valued forest resource systems. One approach might be thought of as land use planning and the other might be thought of as managing joint products.

Today, the land use planning approach dominates decision making. Here the content of management information is typically insufficient to allow true optimization. Unable to fully specify dynamic joint production functions, the land use planning approach usually identifies a finite and rather small number of discrete alternatives and analyzes them in terms of the feasibility and suitability of the land to carry various use levels. Often non market values of resource service flows are excluded from the analysis as well. Timber production is usually constrained in order to fulfill other multiple use responsibilities. One is often left with the impression in the land use planning approach that the most socially efficient combination of resource uses is missed when the planning alternatives are defined. The notion of marginal changes in production when areas as large as a single national forest are "planned" along four or five alternatives is virtually meaningless.

The joint products approach is by far the more appealing approach from a theoretical perspective. While there has been some excellent preliminary work in the area by Weeks (1975) with a sustained yield constraint, Hartman (1976), Bowes and Krutilla (1979), Calish, Fight and Teeguarden (1978), and Convery (1977). Only Howard's (1977) effort to develop actual joint products production functions which comes to mind. In fact, the lack of biotechnological tradeoff knowledge is so severe, that one is left with the fear that the externalities are at times more perceived than real.

The use of the term "uncompensated benefits and costs" is intended to suggest a more general meaning than the term "externality". The optimum management for multiple uses is confronted with several related analytical problems. First there are problems in defining property rights. An example might be that owners of big game winter range often differ from owners of land where game investigate several additional investment alternatives. For example, where partial cutting is considered, the opportunity costs of reserve volumes left on the site is determinant. Tradeoffs between the various reproduction systems and the expected costs of reforestation associated with the reproduction alternatives can be estimated. Merzenich (1979) has used the equations in order to calculate investment returns based upon intrinsic site characteristics. The equation shows for example, that where sites are cable yarded as opposed to tractor skidded (due to steepness of slope) or where certain habitat types produce certain tree species that have different relative values than others, net stumpage value and resulting return on investment will be affected.

Many silvicultural interventions affect tree diameter as well as volume per acre. Hence estimating timber value as a result of both diameter and volume growth allows better estimates of investment return.

Uneven aged management is the result of manipulating stands via selection cutting systems. In an uneven aged stand of timber, there are essentially four interelated decisions to be made (Marquis, 1978). One must determine the optimum residual volume in the stand after selection cutting or the optimum reserve physical capital stock. In addition, one must determine how the residual capital is to be allocated to different diameter sizes including the largest diameter tree (optimum value of reserve capital stock). On the basis of optimum allocation of residual growing stock, growth predictions are made. Finally, the cutting cycle (analogous to rotation length) is made. In effect, these four interelated decisions are like a set of simultaneous equations. An accurate estimate of tree value based on diameter and volume removed will increase the accuracy of estimated return on capital with uneven age management.

In summary, there are several avenues in the area of silvicultural economics that warrant further investigation. While all economists recognize the implicit costs associated with prolonged timber production periods, too little attention has been focused on fine tuning the investment decisions particularly in the area of investments which alter stand quality in addition to stand quantity.

A CLOSING NOTE

Hopefully, this research presents a broader perspective and integration of the concepts of financial maturity with variable real investment and private supply in the long run. It should also serve to identify areas that warrant further research investigation. To both foresters and conservationists, the proof of a positive supply function without old growth timber and technological change, should be important. For economists, the merging of investment and current output decisions via marginal analysis and the theory of optimal timing is a most fruitful extension of contemporary resource management criteria with significant policy implications.

Appendix A

This appendix refers to chapter 2.

APPENDIX A.1 Determination of Single vs. Infinite Series Rotation Age with a Fixed Yield Function

Let: The quantity overtime be expressed as:

$V = A - Be^{-bt}$ (volume per acre) t = time

P = product price net of harvesting and hauling costs (i.e. independent of total vol/ac.)

i = discount rate

q_0 = fixed planting cost (with instantaneous regeneration)

For the infinite series of harvests, the objective function is the Maximization of wealth (W).

$$\text{Max } W = \frac{PV - q_0 e^{it}}{e^{it} - 1}$$

The necessary conditions for the Maximization of wealth with the time of maturity variable is that

$$P\,\partial V/\partial t = \frac{i[PV - q_0]}{1 - e^{it}}$$ see (Hirschleiffer, 1970/p.89)

$$P\,\partial V/\partial t = \frac{i[PV - q_0 e^{it}]}{[e^{it} - 1]} + iPV$$

In equation (1') iPV is the opportunity cost of holding the standing timber an additional instant in time. The explanation of $i\left[\frac{PV - q_0 e^{it}}{e^{it} - 1}\right]$ is the flow value of the perpetual of harvests or the soil rent.
Suppose the land has some other valued non timber use.

Let H = the flow value in alternative use.

If the owner is indifferent between allocating land to timber and the alternative use, under the assumption of wealth maximization,

$$(2) \quad H = i\,\frac{PV - q_0 e^{it}}{(e^{it} - 1)}$$

If the price (P) increases to P*, the owner prefers timber production to the alternative use

$$(3) \quad H < i\,\frac{P^*V - q_0 e^{it}}{e^{it} - 1}$$

ie. the land value increase even at the original optimum maturity.

At the original price level P, a unit of land was marginal. The owner was indifferent in allocating his land to timber as opposed to non timber use. As the price increased, the land became non marginal, the owner preferred timber production to alternative use with an unchanged alternative value H. If in the original case H had increased to H* where

$H^* = i\,\frac{[PV - q_0 e^{it}]}{(e^{it} - 1)}$, the owner would allocate land to its alternative use. The timber production use of land would become submarginal. For any stands of timber that were approaching maturity, an increase in H to H* indicated that the stands would be harvested at a point where

$$(4) \quad P\partial V/\partial t = H^* + iPV$$

the opportunity cost of non timber production would result in a single terminal harvest rotation. The more common approach to single series rotation is under circumstances in which the present value of the growing space in timber and non timber use is zero.

APPENDIX A.2 Conditions for Maximization

1st order conditions (necessary)

(5a) $\partial W/\partial X = P\partial V/\partial X - we^{it} = 0$

(5b) $\partial W/\partial t = P\partial V/\partial t - i\frac{[PV - wX]}{[1 - e^{it}]} = 0$

2nd order conditions (sufficient)
The second order conditions formed by the 2nd order derivatives of the objective function form a (2x2) Hessian determinant which must be positive (even order determinant) in order for the solution to be a maximum.

(5a') $P\partial^2V/\partial X^2 + P\frac{\partial^2 V}{\partial X\partial t} - iwe^{it}$

(5b') $P\partial^2V/\partial X\partial t - iwe^{it} + P\partial^2V/\partial t^2 - iP\partial V/\partial t$

$$A = P^2 \begin{bmatrix} -X^2Ke^{-X}Be^{-bt} & - \quad XKe^{-X}Be^{-bt}(b+i) \\ -XKe^{-X}Be^{-bt}(b+i) & (i+Ke^{-X})(-bBe^{-bt})(b+i) \end{bmatrix}$$

$$A = P^2X^2Ke^{-X}(Be^{-bt})^2(b+i) \begin{bmatrix} (-1) & Ke^{-X} \\ -(b+i) & -b(1+Ke^{-X}) \end{bmatrix}$$

$$\therefore \quad A > 0 \iff (b-iKe^{-X}) > 0$$

APPENDIX A.3 Effect of Δ P on Decision Variables Planting and Maturity: Substitutes)

$$\begin{bmatrix} \text{Determinant A} \\ \text{From Appendix A.2} \end{bmatrix} \begin{bmatrix} \partial X/\partial P \\ \partial t/\partial P \end{bmatrix} = \begin{bmatrix} -\partial V/\partial X \\ -\partial V/\partial t + \frac{iV}{(i-e^{it})} \end{bmatrix}$$

Appendix A.3 (continued)

where $-\partial V/\partial X = -XKe^{-X}Be^{-bt} < 0$

$$-\partial V/\partial t + i\frac{V}{(1-e^{it})} > 0 \text{ from equation (7b)}$$

$$P\frac{\partial V}{\partial t} = \frac{i(PV-wX)}{(1-e^{it})} \Rightarrow P(\partial V/\partial t - \frac{iV}{(1-e^{it})} = -wX/(1-e^{it})$$

$$\Rightarrow -\partial V/\partial t + \frac{iV}{(1-e^{it})} = \frac{wX}{(1-e^{it})} \frac{1}{P} > 0$$

$$\frac{\partial X}{\partial P} = \frac{\begin{bmatrix} -\frac{\partial V}{\partial X} & a_{12} < 0 \\ -\frac{\partial V}{\partial t} + \frac{iV}{(1-e^{it})} & a_{22} < 0 \end{bmatrix}}{[A]} > 0 \text{ by Cramer's Rule}$$

and

$$\frac{\partial t}{\partial P} = \frac{\begin{bmatrix} a_{12} & -\partial V/\partial X \\ a_{21} & -\frac{\partial V}{\partial t} + \frac{iV}{(1-e^{it})} \end{bmatrix}}{[A]} < 0$$

APPENDIX A.4 Effect of a Change in Factor Cost of Planting (w) on Optimum Maturity (t_m) and Planting (X) [Time and Planting = Substitutes]

$$\begin{bmatrix} & \\ & A & \\ & \end{bmatrix} \begin{bmatrix} \frac{\partial X}{\partial w} \\ \\ \frac{\partial t}{\partial w} \end{bmatrix} = \begin{bmatrix} e^{it} \\ \\ \frac{-X}{(1-e^{it})} \end{bmatrix}$$

Appendix A.4 (continued)

$$\frac{\partial X}{\partial w} = \frac{\begin{bmatrix} e^{it} & a_{12} \\ \frac{-iX}{(1-e^{it})} & a_{22} \end{bmatrix}}{[A]} < 0 \quad \text{by Cramer's Rule}$$

and

$$\frac{\partial t}{\partial w} = \frac{\begin{bmatrix} a_{11} & e^{it} \\ a_{21} & \frac{-iX}{(1-e^{it})} \end{bmatrix}}{[A]} > 0$$

APPENDIX A.5 Effect of Δ i on Maturity and Planting (t and X; Substitutes)

$$\begin{bmatrix} A \end{bmatrix} \begin{bmatrix} \frac{\partial X}{\partial i} \\ \frac{\partial t}{\partial i} \end{bmatrix} = \begin{bmatrix} t_m we^{it} = Z_1 > 0 \\ \frac{PV-wX}{(1-e^{it})} \quad 1 - \frac{t_m e^{it}}{(1-e^{it})} = Z_2 < 0 \end{bmatrix}$$

$$\frac{\partial X}{\partial i} = \frac{\begin{bmatrix} Z_1 & a_{12} \\ Z_2 & a_{22} \end{bmatrix}}{[A] > 0} = \frac{Z_1 a_{22} - Z_2 a_{12}}{[A]} < 0$$

$$\frac{\partial t}{\partial i} = \frac{\begin{bmatrix} a_{11} < 0 & Z_1 > 0 \\ a_{21} < 0 & Z_2 < 0 \end{bmatrix}}{[A]} = \frac{a_{11} Z_2 - a_{21} Z_1}{[A]} > 0$$

APPENDIX A.6 Effect of ΔP on Volume, Rotation age, and Flow per Unit of Time

from Chapter II

$$(10)\ \partial V/\partial P = \frac{\partial V}{\partial X}\frac{\partial X}{\partial P} + \frac{\partial V}{\partial t}\frac{\partial t}{\partial P}$$

$$\partial V/\partial X \frac{\partial X}{\partial P} = \frac{\partial V}{\partial X} \frac{\begin{bmatrix} \frac{-\partial V}{\partial X} & a_{12} \\ \frac{-\partial V}{\partial t} + \frac{iV}{(1-e^{it})} & a_{22} \end{bmatrix}}{[A]}$$

$$= \frac{XKe^{-X}Be^{-bt}}{[A]}\{(-XKe^{-X}Be^{-bt})(P(1+Ke^{-X})(bBe^{-bt})(-b-i)$$

$$- (PXKe^{-X}Be^{-bt})(-b-i)\ (-bBe^{-bt})(1+Ke^{-X}) + \frac{iV}{(1-e^{it})}\}$$

$$\frac{\partial V}{\partial t}\frac{\partial t}{\partial P} = \frac{\partial V}{\partial t} \frac{\begin{bmatrix} a_{11} & -\partial V/\partial X \\ a_{12} & \frac{-\partial V}{\partial t} + \frac{iV}{(1-e^{it})} \end{bmatrix}}{[A]}$$

$$= \frac{bBe^{-bt}(1+Ke^{-X})}{[A]}[-PX^2Ke^{-X}Be^{-bt}\ [-bBe^{-bt}(1+Ke^{-X})$$

$$+ iV(\frac{1}{1-e^{it}})] - [-XKe^{-X}Be^{-bt}(PXKe^{-X}Be^{-bt})(-b-i)]]$$

$$\frac{\partial V}{\partial P} = \frac{(XKe^{-X}Be^{-bt})^2 P(1-Ke^{-X})(bBe^{-bt})(b+i)}{[A]} -$$

$$\frac{P(XKe^{-X}Be^{-bt})^2(b+i)(bBe^{-bt})(1+Ke^{-X})}{[A]} +$$

$$\frac{P(XKe^{-X}Be^{-bt})^2(b+i)(iV)\left(\frac{1}{1-e}it\right)}{[A]} + \frac{(bBe^{-bt})^2(1+Ke^{-X})^2 PX^2Ke^{-X}Be^{-bt}}{[A]}$$

$$- \frac{P(XBe^{-bt})^2 b(1-Ke^{-X})(iV)\frac{1}{(1-e^{it})}}{[A]} - \frac{P(XKe^{-X}Be^{-bt})^2(b+i)bBe^{-bt}(1+Ke^{-x})}{[A]}$$

$$\frac{\partial V}{\partial P} = \frac{P(XKe^{-X})^2 Ke^{-X}(b-iKe^{-X}) - iV\left(\frac{1}{1-e}it\right) + bBe^{-bt}(1+Ke^{-X})}{P^2X^2Ke^{-X}(bBe^{-bt})^2(b+i)(b-iKe^{-X})}$$

$$\frac{\partial V}{\partial P} = \frac{-iV\left(\frac{1}{1-e}it\right) + bdBe^{-bt}(1+Ke^{-X})}{P(b+i)} < 0$$

Note: Numerator 0 from equation (7b)

From Figure 2.2 in text;

Determine $\frac{\Delta CX}{\Delta t_m} = \frac{\partial V/\partial P < 0}{\partial t/\partial P < 0}$ — $\partial V/\partial P < 0$ from above; $\partial t/\partial P < 0$ from Appendix A.3

$$\frac{\left(\dfrac{-iV\left(\frac{1}{1-e}it\right) + bBe^{-bt}(1+Ke^{-X})}{P(b+i)}\right)}{\dfrac{-PX^2Ke^{-X}Be^{-bt}\left[-bBe^{-bt}+iV\frac{1}{(1-e^{-it})}\right] - P(XKe^{-X}Be^{-bt})(XKe^{-X}Be^{-bt})(b+i)}{P^2(XBe^{-bt})^2(Ke^{-X})(b+i)(b-iKe^{-X})}}$$

$$= \frac{(b-iKe^{-X})[bBe^{-bt}(1+Ke^{-X})-iV(\frac{0}{1-e^{it}})]}{(bBe^{-bt})(1+Ke^{-X})-iV\left[\frac{1}{(1-e^{it})}-Ke^{-X}(b+i)\right]}$$

Note: $(b-iKe^{-X}) < 1$ Appendix A.1

and $0 > bBe^{-bt}(1+Ke^{-X})\ \frac{iV}{1-e^{it}})$

$$\therefore \quad 0 < \frac{\partial V/\partial P}{\partial t/\partial P} = \frac{\Delta CX}{\Delta t_m} < 1$$

Therefore if OC has a slope ≥ 1 then MAI @ X > MAI @ C

IE. $\frac{\Delta CX}{\Delta t} - 1 < 0$

The result of this proof is that if the original equilibrium mean annual increment (V/t) is greater than the trivial level of one unit of volume per acre, an increase in stumpage price will shorten optimum harvest age, increase optimum planting, reduce volume harvested but increase per acre flow.

APPENDIX A.7 Determination of Δ Interest Rate (i) on Volume-Rotation Age Expansion Path (X and t: substitutes)

$$\partial V/\partial i = \frac{\partial V}{\partial X}\frac{\partial X}{\partial i} + \frac{\partial V}{\partial t}\frac{\partial t}{\partial i}$$

$$\frac{\partial V}{\partial i} = XKe^{-X}Be^{-bt}\ \frac{[Be^{-bt}(b+i)[Z_1(1+Ke^{-X})(-b)-XKe^{-X}(Z_2)}{[A]}$$

$$+ \frac{bBe^{-bt}(1+Ke^{-X})[XKe^{-X}Be^{-bt}\ [-X(Z_2)-Z_1(-b-i)]]}{[A]}$$

from Appendix A.5

$$\frac{\partial V}{\partial i} = \frac{XKe^{-X}(Be^{-bt})^2 Z_2[(-b-i)(XKe^{-X})+b(1+Ke^{-X})]}{[A]} < 0$$

$\therefore$ producing less volume over longer period of time and flow per acre is an inverse function of (i)

Note: $Z_1 = tse^{it} > 0$

$$Z_2 = \frac{PV-wX}{(1-e^{it})}\left[1 - \frac{te^{it}}{(1-e^{it})}\right] < 0$$

See Appendix A.5

APPENDIX A.8 Analysis of Production Opportunity Set where(t) and (X) are Decision Compliments

Production Function: $[V = (A-Be^{-bt})(1-Ke^{-X})]$

$$\text{MAX } W' = \frac{PV-wXe^{it}}{e^{it} - 1}$$

First order conditions

$$\frac{\partial W'}{\partial X} = PXKe^{-X}(A-Be^{-bt}) - we^{it} = 0$$

$$\frac{\partial W'}{\partial t} = PbBe^{-bt}(1-Ke^{-X}) - i\frac{PV'-wX}{(1-e^{it})} = 0$$

Second order conditions

$$\begin{bmatrix} P(-X^2Ke^{-X}(A-Be^{-bt}) & + P\ XKe^{-X}bBe^{-bt} - iwe^{it} \\ P(XKe^{-X}bBe^{-bt} - iwe^{it} & + P(-bBe^{-bt}(1-ke^{-X}) - PibBe^{-bt}(1+Ke^{-X}) \end{bmatrix}$$

where $iwe^{it} = iPXKe^{-X}(A-Be^{-bt})$

$$B = P^2 \begin{bmatrix} -X^2Ke^{-X}(A-Be^{-bt}) & XKe^{-X}bBe^{-bt} - i(XKe^{-X}(A-Be^{-bt}) \\ XKe^{-X}bBe^{-bt} - i(XKe^{-X}(A-Be^{-bt}) & (1-Ke^{-X})(bBe^{-bt})(-b-i) \end{bmatrix}$$

$$b_{12} = b_{21} = XKe^{-X}\,[bBe^{-bt} - i(A-Be^{-bt})]$$

where bBe^{-bt} = rate of growth at maturity of fully stocked stand

$i(A-Be^{-bt})$ = interest rate times total volume of fully stocked stand

$P(bBe^{-bt} - i(A-Be^{-bt}) > 0$ at maturity (see Gaffney, 1957)

We shall simply indicate that the sufficient conditions hold via the convexity of the opportunity set.

IE.

$$B = \begin{vmatrix} \frac{\partial^2 W}{\partial X^2} < 0 & \frac{\partial^2 W}{\partial X \partial t} > 0 \\ \frac{\partial^2 W}{\partial t \partial X} > 0 & \frac{\partial^2 W}{\partial^2 t} < 0 \end{vmatrix} > 0$$

Determining the sign of $\frac{\partial X}{\partial P}$ and $\frac{\partial t}{\partial P}$

$$\begin{bmatrix} B \end{bmatrix} \begin{bmatrix} \frac{\partial X}{\partial P} \\ \frac{\partial t}{\partial P} \end{bmatrix} = \begin{bmatrix} -\frac{\partial V}{\partial X} < 0 \\ \frac{\partial V}{\partial t} + \frac{iV}{(1-e^{it})} \end{bmatrix}$$

and $-i\frac{\partial V}{\partial t} + \frac{iV}{(1-e^{it})}$ 0 from (7b)

APPENDIX A.8 (continued)

$$\frac{\partial X}{\partial P} = \frac{\begin{bmatrix} -XKe^{-X}(A-Be^{-bt}) & PXKe^{-X}(bBe^{-bt} \ (-i(A-Be^{-bt}) \\ -bBe^{-bt}(1-Ke^{-X}) + \frac{i[A-Be^{-bt}](1-Ke^{-X})}{1-e^{it}} & P(bBe^{-bt})(1-Ke^{-X})(-b-i) \end{bmatrix}}{[B]}$$

Let $(A-Be^{-bt}) = y$

$$\frac{\partial X}{\partial P} = \frac{PXKe^{-X}(1-Ke^{-X})}{[B]}$$

$$\begin{bmatrix} -y & (+bBe^{-bt}-iy) \\ -bBe^{-bt}(1-Ke^{-X}) + \frac{i(y)(1-Ke^{-X})}{(1-e^{it})} & bBe^{-bt}(-b-i) \end{bmatrix} =$$

$$\frac{PXKe^{-X}(1-Ke^{-X})}{[B]}\left(\frac{iy^2}{1-e^{it})} + (bBe^{-bt})^2 + X(bBe^{-bt})[b+i-i]\right)$$

$$\therefore \frac{\partial X}{\partial P} > 0$$

$$\frac{\partial t}{\partial P} = \frac{\begin{bmatrix} P(-X^2Ke^{-X})(y) & -XKE^{-X}(y) \\ PXKe^{-X}(bBe^{-bt}-iX) & (1-Ke^{-X})bBe^{-bt} + \frac{iy}{(1-e^{it})} \end{bmatrix}}{[B]} =$$

$$PX^2Ke^{-X}X \frac{\begin{bmatrix} (-1) & (-1) \\ Ke^{-X}(bBe^{-bt}-iy) & (1-Ke^{-X})(-bBe^{-bt} + \frac{iy}{(1-e^{it})} \end{bmatrix}}{[B]}$$

[let $g = bBe^{-bt}$]

$$= \frac{PX^3Ke^{-X}}{[B]} \left[\left[g \frac{-iy}{(1-e^{it})} \right] (1-Ke^{-X}) + Ke^{-X}(g-iy) \right]$$

$$\frac{\partial t}{\partial P} = \left[-g + \frac{iy}{(1-e^{it})} - Ke^{-X} - \frac{Ke^{-X}iy}{(1-e^{-it})} \right] > 0$$

because $g - \frac{iy}{(1-e^{it})} > 0$ from 7b

$$\frac{\partial t}{\partial P} > 0$$

Note: $\frac{\partial V}{\partial P} = \frac{\partial V}{\partial X}\frac{\partial X}{\partial P} + \frac{\partial V}{\partial t}\frac{\partial t}{\partial P} > 0 \qquad \frac{\frac{\partial V}{\partial P}}{\frac{\partial t}{\partial P}} \qquad \frac{\frac{\partial V}{\partial X}\frac{\partial X}{\partial P}}{\frac{\partial t}{\partial P}} + \frac{\partial V}{\partial t}$

∴ an increase in price increases $\frac{V}{t}$ the slope of the volume rotation age expansion path is greater than the growth rate

Direction of Change in Optimum Planting (X) and Harvest Age (t) Resulting from Change in (i)

$$\begin{bmatrix} B \end{bmatrix} \begin{bmatrix} \frac{\partial X}{\partial i} \\ \frac{\partial t}{\partial i} \end{bmatrix} = \begin{bmatrix} twe^{it} > 0 \\ \frac{PV-wX}{1-e^{it})} \quad 1 - \frac{ite^{-it}}{(1-e^{it})} > 0 \end{bmatrix}$$

$$\frac{\partial t}{\partial i} = \frac{\begin{bmatrix} b_{11} < 0 & twe^{it} \\ b_{21} > 0 \quad \frac{PV-wX}{(1-e^{-it})} & 1 - \frac{ite^{-it}}{(1-e^{-it})} > 0 \end{bmatrix}}{[B]}$$

$\frac{\partial t}{\partial i} < 0 \qquad \frac{\partial X}{\partial i} < 0$ because X, t are compliments.

$b_{12} = b_{21} > 0$ => compliments.

$$\begin{bmatrix} B \end{bmatrix} \begin{bmatrix} \frac{\partial X}{\partial w} \\ \frac{\partial t}{w} \end{bmatrix} = \begin{bmatrix} e^{it} > 0 \\ \frac{-iX}{(1-e^{-it})} < 0 \end{bmatrix}$$

Note: $e^{it} = \frac{P\frac{\partial V}{\partial X}}{w}$;

$$\frac{-iX}{(1-e^{-it})} = \frac{P\left[\frac{\partial V}{\partial t} - \frac{iV}{(1-e^{-it})}\right]}{w}$$

$$\frac{\partial t}{\partial w} = \frac{\begin{bmatrix} -PX^2Ke^{-X}(y) & \frac{PXKe^{-X}(y)}{w} \\ PXKe^{-X}(g-iy) & \frac{P(1-Ke^{-X})\left[g - \frac{iy}{(1-e^{-it})}\right]}{w} \end{bmatrix}}{[B]}$$

Let: $y = (A-B^{-bt})$

$g = bBe^{-bt}$

$$= \frac{PX^2Ke^{-X}\frac{1}{w}(y)}{[B]} \begin{bmatrix} -1 & +1 \\ Ke^{-X}(g-iy) & (1-Ke^{-X})\left[g - \frac{iy}{(1-e^{-it})}\right] \end{bmatrix}$$

$$\frac{\partial t}{\partial w} = \frac{PX^2Ke^{-X}\frac{1}{w}(y)}{[B]} \left(\underbrace{g + \frac{iy}{(1-e^{-it})}}_{(<0)} \right) + Ke^{-X}\left(\underbrace{iy - \frac{iy}{(1-e^{-it})}}_{(<0)} \right)$$

$\therefore \quad \frac{\partial t}{\partial w} < 0 \qquad \frac{\partial X}{\partial w} < 0$ (complimentarity between (X) and (t))

APPENDIX A.9 Changes in P with $E(\frac{\dot{P}}{P})$ Constant (Maturity (t) and Planting (X): Substitutes)

$$W = \frac{PV - wXe^{it}}{(e^{it} - 1)} \quad \text{from before}$$

Let $E(\frac{\dot{P}}{P})$= be expected time date of change of price where $\dot{P} = \frac{dP}{dt}$

such that: $0 < E(\dot{P}/P) < i$

and: $i - E(\dot{P}/P) = r$

Wealth equation now becomes

$$W = \frac{PV}{e^{rt}-1} - \frac{wXe^{it}}{(e^{it}-1)} \qquad P = \text{today's price}$$

and $V = (X(P,x,i,E(\frac{\dot{P}}{P}); t(P,w,i,E(\frac{\dot{P}}{P}))$

FIRST ORDER CONDITIONS

$$\frac{\partial W}{\partial X} = \frac{P\frac{\partial V}{\partial X}}{e^{rt}-1} - \frac{we^{it}}{e^{it}-1} = 0$$

$$\frac{\partial W}{\partial t} = \frac{P\frac{\partial V}{\partial t} - re^{rt}PV}{(e^{rt}-1)^2} = \frac{ie^{it}w(e^{it}-1) - ie^{it}we^{it}}{(e^{it}-1)^2} = 0$$

2nd order conditions form (2x2) determinant C where,

$$C_{12} = C_{21} = \frac{P\partial^2 t}{\partial X \partial t}(e^{it}-1) + ie^{it}P\frac{\partial V}{\partial t} - iwe^{it}(e^{rt}-1) - re^{rt}we^{it}$$

$C_{12} = C_{21} < 0$

hence substitutes and

$\frac{\partial X}{\partial P} > 0; \frac{\partial t}{\partial P} < 0$ as before

APPENDIX A.10 On Short Term Speculation

Where today's gross stumpage prices is

$$P_{g_t} > E(P_{g_t})$$

A stand which is mature under conditions where

$$P_{g_{t_0}} = E(P_{g_{t_1}})$$

is over mature where today's price exceeds the expected price in the next period

IE.

$$\left[E(P_{g_{t_1}} - C_H) \cdot E(V_{t_1})\right] \left[(P_{g_{t_o}} - C_H)V_{t_0}\right] \bar{a} + i(P_{g_{t_o}} - C_H)V_t$$

$\bar{a}$ = expected bare land value and constant with term-term price fluctuation.

Bibliography

Alberts, William. 1962. Business Cycles, Residential Construction Cycles, and the Mortgage Market. The Journal of Political Economy LXXX, 3: 263-281.

Alchian, Armen. 1959. Costs and Outputs in The Allocation of Economic Resources by Moses Abramovitz and others. Stanford University Press. Palo Alto. 23-40.

Allen, R. G. D. 1938. Mathematical Analysis for Economists. St. Martin's Press. New York. 548 p.

Arrow, Kenneth. 1951. Social Choice and Individual Values. John Wiley and Sons, Inc. New York. 99 p.

Baden, John and Richard Stroup. 1973. Externality, Property Rights and the Management of our National Forests. Journal of Law & Econ., 16.

Bailey, Martin J. 1971. National Income and the Price Level: A Study in Macroeconomic Theory. Sec. Ed. McGraw-Hill Book Co. New York. 278 p.

Bator, Francis M. 1957. The Simple Analytics of Welfare Maximization. American Economic Review. Reprinted in Readings in Microeconomics. Breit, William and Harold M. Hockman eds. 1968. Holt, Reinhart and Wilson, Inc. New York. 497 p.

Bowes, Michael D. and John V. Krutilla. 1979. Multiple Use Forestry and the Economics of the Multiproduct Firm. Mimeo. Resources for the Future. Washington, D.C. 52 pp.

Buchanan, James M. 1968. The Demand and Supply of Public Goods. Rand McNally and Company. Chicago. 214 p.

Bueter, John and James O. Arney. 1972. Lump Sum Bidding on Federal Lands in Western Oregon. Research paper 12. Oregon State University, Corvallis.

Burt, Oscar. 1965. Optimal Replacement Under Risk. Journal of Farm Econ. 47: 324-346.

Calish, Steven, Roger D. Fight and Dennis E. Teeguarden. 1978. How Do Nontimber Values Affect Douglas-Fir Rotations. Journal of Forestry. 76. 217-221.

Chisholm, Anthony H. 1975. Income Taxes and Investment Decisions: The Long Life Appreciating Asset Case. Economic Inquiry. XII: 565-578.

Clawson, Marion. 1975. Forest For Whom and For What. John Hopkins Univ. Press. Baltimore. 175 pp.

Clawson, Marion. 1976. The Economics of National Forest Management. RFF Working Paper EN-6. Washington, D. C. 117 pp.

Coase, R. H. 1937. The Nature of Firm. Economics IV: 308405. Reprintd in A. E. A. Readings in Price Theory. 1952. Selected by a Committee of The American Economics Association. Richard D. Irwin, Inc. Chicago. 568 p.

Convery, Frank J. 1977. Land and Multiple Use. In Research in Forest Economics and Forest Policy. M. Clawson, ed. Research Paper R-3. Resources for the Future, Washington, D. C.

Davis, Kenneth P. 1966. Forest Management: Regulation and Valuation. Sec. Ed. McGraw-Hill Book Company. New York 519 p.

Dorfman, Robert. 1953. Mathematical or "Linear Programming": A Nonmathematical Exposition. American Economic Review. XLIII, 5: 797-825. Reprinted in Reading in Microeconomics. Breit, William and Harold M. Hockman eds. 1968. Holt, Rinehart, and Winsoton, Inc. New York 697 p.

Dowdle, Barney. 1962. Investment Theory and Forest Management Planning. Yale University School of Forestry Bulletin. 67. New Haven 63 p.

Downs, Anthony. 1974. The Successes and Failures of Public Housing Policy. The Public Interest. 34 (Winter): 124145.

Duerr, William A. 1960. Fundamentals of Forestry Economics. McGraw-Hill Book Company. New York. 597 p.

Faustman, Martin. 1849. Calculation of the Value Which Forest Land and Immature Stands Possess for Forestry. In Martin Faustman and the Evolution of Discounted CashFlow. 1968. Tr. W. Linnard. Ed. M. Gane, Commonwealth Forestry Institute. No. 42. University of Oxford.

Ferguson, C. E. 1969. Microeconomic Theory. Rev. Ed. Richard D. Irwin, Inc. Homewood, Illinois. 521 p.

Fisher, Irving. 1930. The Theory of Interest. Reprint 1965. Augustus M. Kelley, Bookseller. New York. 566 p.

Friedman, Milton. 1935. Essays in Positive Economics. The University of Chicago Press. Chicago. Reprinted in Readings in Microeconomics. Breit, William and Harold M. Hockman eds. 1968. Holt, Rinehart, and Winston, Inc. New York. 23-59.

Friedman, Milton. 1962. Price Theory. Rev. Ed. Aldine Publishing Company. Chicago. 285 p.

Gaffney, Mason M. 1957. Concepts of the Financial Maturity of Timber and Other Assets. Agricultural Economics Information Series. No. 62. North Carolina State College. Raleigh.

Gaffney, Mason, M. 1967. Tax Induced Slow Turnover of Capital. Western Economic Journal. V. 4: 308-232.

Goforth, Marcus H. and Thomas J. Mills. 1975. Discounting Perpetually Recurring Payments Under Conditions of Compound Relative Value Increases. Research Note WO-10. U.S.D.A., Washington, D. C. 3 p.

Gregory, G. Robinson. 1972. Forest Resource Economics. Ronald Press. New York. 548 p.

Hartman, Richard. 1976. The Harvesting Decision When the Standing Forest has Value. Economic Inquiry. XIV: 52-58.

Hirschleifer, J. 1970. Investment, Interest and Capital. Prentice-Hall, Inc. Englewood Cliffs, N.J. 320 p.

Henderson, James M. and Richard E. Quandt. 1958. Microeconomic Theory. McGraw-Hill Book Company, Inc. International Student Edition. Kogakuska Company, Ld. Tokyo, 291 p.

Howard, James E. 1977. Marginal Production Costs for Timber, Water, Wildlife and Recreation on a Simulated Georgia Piedmont Watershed. Unpublished Ph. D. thesis. University of Georgia, Athens. 100 pp.

Howe, Charles W. 1979. Natural Resources Economics. John Wiley & Sons. New York. 350 pp.

Husch, Bertram. 1963. Forest Mensuration and Statistics. The Ronald Press Company. New York. 474 p.

Jackson, David H. 1977. Some Structural Components of Contracts as They Relate to Canadian Forest Tenures. Forestry Chronicle. 53. No. 1.

Jackson, David H. and Alan G. McQuillan. Forthcoming. A Technique for Estimating Timber Value Based on Tree Size, Management Variables and Market Conditions. Forest Science.

Johnson, Ronald N. 1977. Competitive Bidding for Federally Owned Timber. Unpublished Ph.D. thesis. University of Washington, Seattle.

Keays, J. L. and J. V. Hutton. 1974. The Effect of Bark on Wood Pulp Yield and Quality and on the Economics of Production. Information Report VP-X-126. Department of Environment, Canadian Forestry Service. Western Forest Products Research Lab., Vancouver, Canada.

Keays, J. L. and T. Szabo. 1974. Forest Yield is Increased by Pulping Tops. Pulp and Paper, March.

Koch, Peter. 1974. Full Tree Utilization of Southern Pine and Hardwoods growing on Southern Pine Sites. In Optimizing the South's Forests. Proceedings of 2nd Regional Technical Conference. Society of American Foresters.

Kramer, Paul J. and Theodore T. Kozlowski. 1960. Physiology of Trees. McGraw-Hill Book Company. New York. 642. p.

Krutilla, John V. and John A. Haigh. 1978. An Integrated Approach to National Forest Management. Environmental Law. 8: 373-415.

Lerner, Abba P. 1944. The Economics of Control. The Macmillan Company. New York. 428 p.

Little, I.M.D. 1949. The Foundations of Welfare Economics. Oxford Economics Papers. N. S. Vol. I, No. 2.

Lutz, F. and V. 1951. The Investment Theory of the Firm. Princeton University Press. 253 p.

Marquis, David A. 1978. Application of Uneven-Aged Silviculture on Public and Private Lands. In Uneven-Aged Silviculture & Management in the United States. U.S.D.A. Forest Service. Timber Management Research. Washington, D.C.

McArdle, R. E., W. H. Meyer and D. Bruce. 1949. The Yield of Douglas-fir in the Pacific Northwest. U.S.D.A. Technical Bulletin. No. 201. Rev.

Merzenich, James P. 1979. Classifying Forest Land Based Upon Its Timber Management Investment Potential: A Case Study of the Lolo National Forest. Montana Forest Conservation Experiment Station. Bulletin 42. Missoula. 146 pp.

Mishan, E. J. 1964. Welfare Economics: Five Introductory Essays. Random House, New York. 229 p.

Musgrave, Richard A. 1959. The Theory of Public Finance. McGraw-Hill Book Company, Inc. New York. 628 p.

Nautiyal, J. c. and D. V. Love. 1971. Some Economic Implications of Charging Stumpage. Forestry Chronicle. 47: 1-4.

Pinchot, Gifford. 1910. The Fight for Conservation. Doubleday Page and Company. Reprinted, 1967. The University of Washington Press. Seattle. 152 p.

Quirk, James P. 1976. Intermediate Economics. Science Research Associates, Inc. Chicago. 359 pp.

Rao, Potluri and Rober LeRoy Miller. 1971. Applied Econometrics. Wadsworth Publishing Company, Inc. Belmont, California. 235 p.

Robinson, Joan. 1941. Rising Supply Price. Economica. New Series VIII: 1-8. Reprinted in A. E. A. Readings in Price Theory. 1952. George J. Stigler and Kenneth E. Boulding eds. Richard D. Irwin, Inc. Homewood, Ill. 568 p.

Samuelson, Paul Anthony. 1947. Foundations of Economic Analysis. Reprinted 1970 by Atheneum. New York. 447 p.

Samuelson, Paul A. 1976. The Economics of Forestry in an Evolving Society. Economic Inquiry. XIV: 466-492.

Samuelson, Paul A. 1955. Diagrammatic Exposition of a Theory of Public Expenditure. Review of Economic and Statistics. XXXVIII. November.

Schreuder, Gerard. 1971. The Simultaneous Determination of Optimal Thinning and Rotation for an Even-Aged Forest. Forest Science 17, 3: 333-339.

Schweitzer, Dennis L., Robert W. Sassaman, and Con H. Schallau. 1972. Allowable Cut Effect: Some Physical and Economic Implications. Journal of Forestry. 70, 7: 415-418.

Scott, Anthony. 1973. Natural Resources: The Economics of Conservation. McClelland and Stewart, Ltd. Toronto. 313 pp.

Seligman, Edwin R. 1921. The Shifting and Incidence of Taxation. 4th ed. Columbia University Press. New York. Chapter 1 reprinted in Readings in the Economics of Taxation. R. A. Musgrave and Carl Shoup eds. 1958. American Economic Association. Richard D. Irwin, Inc. Homewood, Ill.

Spurr, Stephen H. 1964. Forest Ecology. The Ronald Press Company. New York. 352 p.

Smith, David Martyn. 1962. The Practice of Silviculture. John Wiley and Sons, Inc. New York. 578 p.

Stroup, Richard, and John Baden. 1975. Private Rights, Public Choices and the Management of National Forests. Western Wildlands. Autumn.

Teeguarden, Dennis E. 1973. The Allowable Cut Effect: A Comment. Journal of Forestry. 71, 4: 224-226 p.

Thomson, Proctor and Henry N. Goldstein. 1971. Time and Taxes. Western Economic Journal. IX, 1: 27-45.

U. S. Department of Agriculture. 1964. The Principal Laws Relating to Establishment and Administration of the National Forests and to Other Forest Service Activities. Agricultural Handbook No. 20. Revised. Washington, D.C. 127 p.

U. S. Forest Service. Updated regularly. Forest Service Manual. The operating guide for National Forest Service management. U.S.D.A. Washington, D.C.

U.S. Forest Service. 1973. The Outlook for Timber in the United States. Forest Service Report No. 20. U.S. Government Printing Office. Washington, D.C. 367 p.

U.S. 1973. Report of the President's Advisory Panel on Timber and the Environment. U.S.G.P.O. Washington, D.C. 541 pp.

Vaux, Henry. 1973. How Much Land Do We Need for Growing Timber? Journal of Forestry. 71: 399-403.

Waggener, Thomas Runyan. 1969. Some Economic Implications of Sustained Yield as a Forest Regulation Model. Institute of Forest Products. College of Forest Resources. University of Washington. Seattle. 22 p.

Walker, John Lawrence. 1971. An Economic Model for Optimizing the Rate of Timber Harvesting. Unpublished Ph.D. thesis. University of Washington. Seattle.

Walker, John. 1977. Economic Efficiency and the National Forest Management Act of 1976. Journal of Forestry. 75.

Wall, Brian R. 1973. Employment Implications of Projected Timber Output in the Douglas-fir Region, 1970-2000. U.S.D.A. Forest Service Research Note PNW-wll. 11 p.

Weeks, Wm. J. 1975. Efficient Multiple Use Forestry. Presented to Meeting of Western Economic Association. 8 p.

Witte, James G. Jr. 1963. The Microfoundations of the Social Investment Function. The Journal of Political Economy. LXXI, 5: 441-456.